SpringerBriefs in Materials

For further volumes:
http://www.springer.com/series/10111

Donald V. Rosato

Plastics End Use Applications

 Springer

Donald V. Rosato
PlastiSource, Inc.
Concord, MA 01742, USA
gander.psi@verizon.net

ISSN 2192-1091   eISSN 2192-1105
ISBN 978-1-4614-0244-2   e-ISBN 978-1-4614-0245-9
DOI 10.1007/978-1-4614-0245-9
Springer New York Dordrecht Heidelberg London

Library of Congress Control Number: 2011931281

© Springer Science+Business Media, LLC 2011
All rights reserved. This work may not be translated or copied in whole or in part without the written permission of the publisher (Springer Science+Business Media, LLC, 233 Spring Street, New York, NY 10013, USA), except for brief excerpts in connection with reviews or scholarly analysis. Use in connection with any form of information storage and retrieval, electronic adaptation, computer software, or by similar or dissimilar methodology now known or hereafter developed is forbidden.
The use in this publication of trade names, trademarks, service marks, and similar terms, even if they are not identified as such, is not to be taken as an expression of opinion as to whether or not they are subject to proprietary rights.

Printed on acid-free paper

Springer is part of Springer Science+Business Media (www.springer.com)

Preface

This SpringerBrief *Plastics End Use Applications* covers approximately 120+ plastics applications and trends in the major end use areas of electrical and electronic, industrial, transportation, consumer products, medical, emerging, and related subsectors. Plastics play an indispensable role in a wide variety of markets, ranging from packaging and building & construction to transportation, consumer and institutional products, furniture and furnishings, electrical & electronic components, medical and others. This enormous variety of end uses for plastics materials is what gives the industry its remarkably dynamic character. This also offers an exciting and significant challenge to companies, to find the optimum mix of markets and customers to pursue that which fits with the products it makes, the services it offers, and its financial goals.

Plastics End Use Applications provides a simplified, practical, and innovative approach to understanding the design and manufacture of plastic products. This unique review will expand the reader's understanding of plastics technology by defining and focusing on current and future technical trends. Plastics behavior is presented to enhance the capability of fabricating products that meet performance standards, low cost requirements, and profitability targets.

Application examples include toys, medical devices, cars, boats, underwater devices, containers, springs, pipes, buildings, and aircraft. Also covered are behaviors associated with different plastic materials (thermoplastics, elastomers, reinforced plastics) and the many fabricating processes (extrusion, injection molding, blow molding, forming, foaming, reaction injection molding, rotational molding). This material is presented so that both technical and non-technical reader can understand the interrelationships of materials to processes to applications. Also examined are different plastic products and their related critical factors, from meeting performance requirements in different environments, to reducing costs, and targeting for zero defects. Examples used include small to large, and simple to complex shapes. Major specific plastic end-use markets covered include packaging, building and construction, automotive, electrical and electronic, appliance, medical, consumer products, toy, recreation and leisure, furniture, office products, lawn and garden, marine and boat, aerospace, industrial, agriculture, waste management, government, export, and other and emerging.

This SpringerBrief is designed to keep professionals in the global plastics industry abreast of key technical developments, business strategies, and marketing initiatives in plastics and competitive materials from an applications standpoint that impact the sales and usage of plastics. The customized interpretative nature of this SpringerBrief encompasses material, processing, fabricator oriented, major end-use competitive and/or complementary material trend (metal, rubber, paper & wood, glass & ceramic) viewpoints, and displays a unique plastic technology knowledge building focus.

Plastics End Use Applications has been prepared with the awareness that its usefulness will depend on its simplicity and its ability to provide essential information. Preparation for this SpringerBrief drew on information from participating industry personnel, global industry and trade associations, and the author's worldwide personal, industrial, and teaching experiences.

Donald V. Rosato

Contents

Chapter 1
Executive Summary

Keywords Plastic • Electronics • Prototypes • Packaging • Urethane • Composites • Polyphthalamide • Ergonomic • Medical • Electroluminescence

1.1 Overview

1.1.1 Overview

This "Plastics End Use Applications" SpringerBrief covers approximately 120+ plastics applications and trends in the major applications areas of (1) plastics industry fundamentals, (2) electrical and electronic applications, (3) industrial applications, (4) transportation applications, (5) consumer products applications, (6) medical, emerging, and other applications, and (7) plastics applications company sourceguide.

1.1.2 Global Technology Highlights (North America, Europe, Asia)

Let us take a brief tour of the nearly 120+ emerging, global plastics applications highlights covered in the SpringerBrief as follows.

1.1.3 Electrical and Electronic Applications

- Plastic Logic Ltd. (UK), a leading developer of plastic electronics including printed flexible thin film transistor (TFT) arrays, is developing and exploiting a portfolio of intellectual property based on inkjet printing of active electronic

D.V. Rosato, *Plastics End Use Applications*, SpringerBriefs in Materials,
DOI 10.1007/978-1-4614-0245-9_1, © Springer Science+Business Media, LLC 2011

circuits using advanced plastic materials to form thin film transistors that can be used in active matrix back planes that drive displays. Other potential applications include smart labels, smart packaging, and radio frequency identification (RFID) devices. Plastics electronics can be produced directly at high speed from CAD data to large, flexible surfaces using ink jet printing equipment rather than complex photolithography and vacuum systems used to make today's transistors. With the process' low temperatures, substrates used can also be plastic.

- Dyson Ltd. (UK) realized that with most vacuum cleaners the bag quickly clogs with dust as the suction had to pass through the bag, and a clogged bag, even with as little as 10 oz. of dirt, can cut the suction in half so the more a bag is used the less effective it becomes. Entrepreneur James Dyson set out to solve the problem. Five years and 5,000 prototypes later, the world's first cyclonic bagless vacuum arrived. When none of the major manufacturers were interested, Dyson launched his own vacuum cleaner company. Now, Dyson Cyclonic Vacuum, Europe's hottest home appliance, has come to America.

1.1.4 Industrial Applications

- OnTech (U.S.) has one of the most innovative packaging solutions recently introduced as North America's first self-heating container, designed to heat liquid contents such as coffee, tea, cocoa, soups, and alcoholic beverages. Created by OnTech, Inc., the new self-heating container is a safe and easy-to-activate concept that heats up the contents of its package to approximately 145°F within minutes. The company now has 102 approved utility patent claims in the US and patents in 36 other countries, including the UK, China, and Japan covering any multi-chamber plastic product that must be retorted, or sterilized during a heating process. In Europe, some metal containers can self-heat and others have tried a mix of materials to create the same result but with limited commercial results. Metal has a tendency to collapse or crush during the sterilization process, when the container has to be heated to 250°F.
- According to Style Solutions Inc. (U.S.), customized construction products and large builders will drive the growth of urethane millwork in 2005. The firm has created more than 300 customized products in the past year for builders across the US. The building and construction industry has seen unprecedented advances in the acceptance and use of urethane millwork on both exteriors and interiors of commercial and residential buildings. Just 5 years ago, wood moulding profiles was predominate with few engineered plastic ones seen. Today the reverse is true with low-maintenance products, such as urethane, composite, and plastic mouldings prominent in all regions of the country. Urethane millwork, which stands up to high temperatures in the summer and frigid snowy weather in the winter months, is being increasingly used with other products such as doors, windows, and vinyl siding.

1.1.5 Transportation Applications

- Johnson Controls (U.S.) states that a top of the line car is now equipped with an average of ten antennae, and that number is expected to soon increase to as many as 16 as car manufacturers introduce systems for monitoring and automatically regulating the distance from the vehicle in front, and install warning sensors to avoid collisions when reversing. These numerous antennae have been generally spread over the entire bodyshell, presenting a challenge to designers who have to guarantee optimum reception. A new antenna module developed as a joint project by Johnson Controls and Volvo demonstrates a way round the problem. The module accommodates all the antennae and relevant receivers (radio, TV, GPS) in one unit. Made of steel and Durethan BM 130 H2.0 polyamide supplied by Bayer MaterialScience, it is manufactured using the plastic/metal hybrid technology already established for production of vehicle front-ends.
- Scaled Composites (U.S.) and Virgin Atlantic (UK) have unveiled the GlobalFlyer V. The pioneering aircraft, said to be the world's most efficient jet plane, has been piloted in an attempt to achieve the first solo, nonstop around the world flight. The aircraft completed the 80-h voyage on one tank of fuel. Virgin Atlantic is the sponsor the 38.7-foot long, single engine jet, with a 114-foot wingspan. Scaled Composites spent more than 4 years designing and building the aircraft. The company used computer-aided aerodynamic studies and built the structure entirely from ultra-light carbon, Kevlar, glass, and a combination of reinforced epoxy and polyester composite materials.

1.1.6 Consumer Products Applications

- Solvay Advanced Polymers' (U.S., Belgium) Amodel Polyphthalamide (PPA) has broken into the wine industry. Typically found in automotive, electrical, and industrial product applications, Solvay Advanced Polymers LLC's high-performance polymer has found a new use as a threaded anchor for the patented twist-to-uncork wine packaging designed by Gardner Technologies, Inc., a Napa, CA, based manufacturer. The new product carries the MetaCork trade name and was on display at the recent K Fair in Düsseldorf. MetaCork™ consists of a hard plastic capsule with a threaded interior surface, a matching plastic threaded cap, and a natural or synthetic cork fitted with a threaded anchor, made from Amodel PPA, that is screwed into the cork during the bottling process.
- An attractive new chair, part of contract furnishings manufacturer Allsteel Inc.'s (U.S.) 'Get Set' line of office chairs, tables, and accessories, is its Get Set multi-purpose room seating/office side chair. The new chair, a comfortable, flexible solution for learning environments, is designed for both horizontal nesting and vertical stacking (up to four high) for versatile, easy transportation and storage. Though portability is a plus, the benefit of Get Set chairs is the hours of comfortable seating they provide. Unlike non-padded folding or stacking

chairs often used in a learning setting, Get Set chairs focus on comfort, featuring seat cushions and a unique 'flex back,' which has a structure that flexes naturally, to counterbalance body weight and conform to support users of different shapes and sizes. An ergonomic sloping arm option also allows for natural movement of the body. Perforations in the chair back allow air circulation, an important consideration, as well. Durable multi-surface casters make movement easy.

1.1.7 Medical, Emerging, and Other Applications

- Puratech GmbH's (Switzerland) operating microscopes play a crucial role in the operating room. They must deliver top performance in a highly demanding environment and they must be lightweight, always accessible but never in the way during demanding and detailed work. Consequently, these instruments are equipped with highly sophisticated balancing systems that ensure high-precision movement of the complex optical system without manual involvement on the part of the surgeon. The arm which the microscope moves on must therefore be as light as possible. High-performance plastics like the lightweight, robust, Baydur 110 polyurethane system from Bayer MaterialScience AG are ideal for use in this application.
- Bree Collection GmbH (Germany), an international leather and bag specialist, has created an illuminated business handbag using Smart Surface Technology (SST). Bree had been toying with the idea of illuminating the dark insides of handbags for some time, but lacked an elegant solution. The answer is now available as Smart Surface Technology developed by Bayer Polymers, in partnership with Lumitec, a specialist in electroluminescence (EL) and precision electronic components. EL is method of generating light reminiscent of the chemical method employed by fireflies. Engineers are using a film that lights up on application of a voltage to achieve electroluminescence. An advantage of EL is that it does not produce heat. However until now, only flat surfaces of limited size could be achieved. Smart Surface Technology makes it possible for the films to be shaped to illuminate any conceivable geometry. Incorporated in a nonconductive layer, the film in Bree's handbag lights up at the press of a button. Bayer Polymers sees the main application for this technology in the auto industry. Incandescent lamps in cars will become obsolete. Instrument panels will be designed to take up less room and a car's interior headliner will glow in a soft glare-free light.

1.2 Scope and Methodology

1.2.1 Scope

The Plastics End Use Applications scope encompasses emerging technology and trends in the following major and subtopic plastics applications focused areas:

Electrical and electronic applications
- Electrical and electronic introduction
- Fuel cell plates
- PLED flat panel displays
- Flexible printed transistors
- High-performance ignition Coil Bobbins
- Smart wallplates
- Electronic cloth
- Intrinsic EMI shielded products
- Miniaturized DSL transformers
- Electrodynamic loudspeakers
- Visual/tactile electronic enclosures
- High-performance outdoor cable

Office products applications
- Miniaturized fuel cells
- Rollable reading displays
- Digital pens
- Orbital web cameras
- Removable data storage media

Appliance applications
- Appliance applications introduction
- E/E design for recycle appliances
- Dyson cyclone vacuum cleaner
- High-performance light fixtures

Packaging applications
- Packaging applications introduction
- Stretch hood film wrap
- Electronic product codes (RFID)
- Transparent paint cans
- Elastic shrinking stretch film
- Rectangular beverage bottles

Building and construction applications
- Building and construction applications introduction
- Fiber-based composite building products
- Flexible mouldings
- Artistically designed floor coverings
- Sandwich construction plate
- Secure window film

Industrial applications
- Industrial applications introduction
- Energy saving freezer apparatus
- Plastic pallets
- Sterile disposable flasks
- Mechanically efficient microplates
- Aerodynamic truck brackets
- High tech fasteners

Agricultural applications
- Agricultural applications introduction
- Farmland demining equipment
- Engineered horseshoes
- Ag equipment large parts
- High durability irrigation filters
- Ag pesticide recyclable containers

Automotive applications
- Plastics and transportation
- Automotive applications introduction
- Carburetor and air intake system firsts
- LESA part adhesive technology
- Optical glazing
- Class A roof appliques
- Carbon fiber concept cars
- Metal/plastic antenna modules
- High value oil valve covers
- Composite front hood
- Ford's GloCar

Aerospace applications
- Aerospace applications introduction
- Unmanned aerial vehicles
- A380 composite floor panels
- Ultra lightweight global flyer
- Mars tested circuit materials
- Robotic jet bombers
- Parachute safe air transport
- 7E7 dreamliner composites

Marine and boat applications
- Marine and boat applications introduction
- State-of-the-art outboard engines
- High tech pontoon boats
- Subsea high strength moorings
- Small craft rotomolding
- Corrosion resistant pier sleeves
- Light weight engineered dock pilings

Consumer products applications
- Consumer products and the consumer
- A new wine cork twist
- High tech gloves
- Consumer durable cushioning
- High capacity Blu-Ray DVDs
- Electronic book binding
- EU food compliant containers
- High-temperature cookware
- High tech power shavers
- Blow molded standup tubes

(continued)

(continued)

– Do-it-yourself faucets	– Non-slip garden tools
– Time expired DVDs	– Functional stylized glazing
Toy applications	*Medical applications*
– Toy applications market	– Medical applications introduction
– Rotomolded toys	– High tech IV catheters
– Large-scale building toy	– Life-saving items safely packaged
– Highly precise game joy sticks	– Highly maneuverable surgical microscope
– High tech learning toys	– Surgical microtools
– Import proof seasonal toys	– Micro-sized assay disks
Recreation/leisure applications	– Home healthcare CPAP systems
– Recreation/leisure applications introduction	*Emerging applications*
– Ultralight foldable kayak	– Emerging applications introduction
– Clear corrosion-resistant pool cleaners	– Plastic bone wound repair
– Cross country skiing/skating	– Satellite TV car antennas
– Snow skates	– Luminescent handbags
– Laserline golf tees	– 'Virtual layering' clothing fabrics
– Light, efficient golf club merchandising carts	*Other applications*
Furniture applications	– Government applications introduction
– Furniture applications introduction	– Driver safe lamp posts
– Polyurethane foam quilting replacement	– Ageless underground pipe
– Lightweight composite tables	– Super strong water pipe
– Naturally flexing office chairs	– Composite UCAVs
– Performance enhanced window shades	– Waste management applications introduction
– Renewable resource sleep products	– Recycled PET mine shaft reinforcement
– Environmentally friendly chairs	– Waste carpet fiber products
Lawn and garden applications	– Nylon car parts recycling
– Lawn and garden introduction	– Eco-efficient mechanical recycling
– Novel plastic fencing	– Environmentally marketed irrigation pipe
– Miracle-gro multipurpose pails	

1.2.2   Methodology

The methodology to create Plastics End Use Applications included globally developing and reviewing data, information, and analysis in emerging technology and trends from multiple sources such as:

1. Direct company input from 120+ technology sources as outlined below in the webwatch directory and Chapter eight's applications company sourceguide
2. 10+ major trade shows (i.e., NPE, Fakuma, K Fair, etc.)
3. 5+ major annual trade association meetings (i.e., ANTEC, APME, PPI, etc.)
4. 15+ major trade journals (i.e., Plastics Technology, Kunststoffe Plast Europe, Modern Plastics International, etc.), and
5. 15+ major database (i.e., Derwent World Patents, etc.), reference work (i.e. Plastics Institute of America Plastics, Engineering, Manufacturing and Data Handbook, etc.) and study sources (i.e., Freedonia, etc.).

1.2.3   Plastics End Use Applications Webwatch Directory

A global webwatch directory of 120+ company names and web addresses arranged chronologically from Chaps. 3–7 are summarized for your reference convenience immediately below. Additional contact information for 625 global plastics applications oriented companies is contained in Chap. 8.

3. Electrical and electronics end use applications	
Bulk Molding Compounds Inc.	www.bulkmolding.com
Cambridge Display Technologies Ltd.	www-cdtltd.ellipsismedia.net
Plastic Logic Ltd.	www.plasticlogic.com
MSD Ignitions Div., Autotronic Controls Corp.	www.msdignition.com
Cooper Industries, Inc.	www.cooperindustries.com
University of California at Berkeley	www.eecs.berkeley.edu
DuPont	www.plastics.dupont.com
Epcos AG	www.epcos.com
Cabasse	www.cabasse.com
Inclosia Solutions, Dow Chemical Co.	www.inclosia.com
Teknor Apex Vinyl Division	www.teknorapex.com
MTI MicroFuel Cells Inc.	www.microfuelcell.com
PolymerVision, Philips Technology Incubator	www.hightechcampus.nl
Logitech, Inc.	www.logitech.com
Imation Corp.	www.imation.com
Mitsubishi Electric Co., Ltd.	www.mitsubishi.com
Dyson Ltd.	www.dyson.com
Miele & Cie. KG	www.miele.de
4. Industrial end use applications	
Lachenmeier A/S	www.lachenmeier.com
Wal-Mart Stores, Inc.	www.walmartstores.com
PCC Group	www.plastic-can.com
OnTech, Inc.	www.ontech.com
Easiwrap International Ltd.	www.easiwrap.net
Owens-Illinois, Inc.	www.o-i.com
Kadant Composites, Inc.	www.kadant.com; www.geodeck.com
Style Solutions, Inc.	www.stylesolutionsinc.com
Bayer MaterialScience AG	www.bayermaterialscience.com; www.artwalk-bayer.com
Intelligent Engineering Ltd.	www.ie-sps.com
CPFilms Inc.	www.cpfilms.com
Creative Plastics & Design, Inc.	www.cpdesign-inc.com
Kiga GmbH	www.kiga-gmbh.de
Nalge Nunc International, Inc.	www.nalgenelabware.com
Tecan Group Ltd.	www.tecan.com
RTP Co.	www.rtpcompany.com
Icotec AG	www.icotec.ch
Countermine Technologies AB	www.countermine.se
Senior and Dickson	www.jameg-sprinters.co.uk
Bemis Plastics	www.bemisplastics.com/agriculture

(continued)

(continued)

Irritrol Systems Europe S.r.l.	www.irritrol.it
Ag Container Recycling Council	www.acrecycle.org

5. *Transportation end use applications*

Solvay Advanced Polymers LLC	www.solvayadvancedpolymers.com
Dow Automotive	www.dow.com/automotive
Schefenacker/Freeglass AG	www.schefenacker.com
DuPont Canada Inc.	www.ca.dupont.com
Meridian Auto. Systems, Inc.	www.meridianautosystems.com
Bayer MaterialsScience AG	www.bayermaterialscience.com
Bruss North America	www.brussna.com; www.bruss.de
Stamax BV, SABIC Europe	//polymers.sabic-europe.com
General Motors Corp.	www.gm.com
Ford Motor Co.	www.ford.com
Northrop Grumman	www.northropgrumman.com
MC Gill Corp.	www.mcgillcorp.com
Scaled Composites, LLC	www.scaled.com
DuPont Co.	www.dupont.com/et
Boeing Phantom Works	www.boeing.com/phantom
Ballistic Recovery Systems Inc.	//brsparachutes.com
Boeing Co.	www.boeing.com
Mercury Marine	www.mercurymarine.com
Genmar Holdings Inc.	www.genmar.com
Victrex plc	www.victrex.com
Brigham Young University	www.byu.edu
MFG Construction Products Co.	www.mfgcp.com
ArmorDock	www.armordock.com

6. *Consumer products end use applications*

Ergodyne	www.ergodyne.com
Rogers Corp.	www.rogerscorporation.com
Sony Disc Technology Inc.	www.sdt.sony.co.jp
Sagoma Plastics	www.sagomaplastics.com
Rubbermaid Europe	www.rubbermaid-europe.com
BASF Engineering Plastics	www.basf-ag.de
The Gillette Company	www.gillette.com
Graham Packaging Co.	www.grahampackaging.com
Moen Inc.	www.moen.com
Flexplay Technologies, Inc.	www.flexplay.com
Step2 Co.	www.step2.com
Sagoma Plastics	www.sagomaplastics.com
Logitech Inc.	www.logitech.com
LeapFrog Enterprises, Inc.	www.leapfrog.com
Grand Venture Ltd.	www.grandventure.com
Firstlight Kayaks	www.firstlightkayaks.com
GE Advanced Materials	www.geadvancedmaterials.com
GateSkate, Inc.	www.gateskate.com
Salomon	www.salomonsports.com
Spirit of Golf GmbH	E-mail: info@spog.de

(continued)

(continued)

Meese Orbitron Dunne Co.	www.modroto.com
Bayer MaterialsScience AG	www.bayermaterialscience.com
Allsteel Inc.	www.allsteeloffice.com
Hunter Douglas, Inc.	www.hunterdouglas.com
Pacific Coast Feather Co.	www.pacificcoast.com
Herman Miller Inc.	www.hermanmiller.com
Kroy Building Products, Inc.	www.kroybp.com
Unimark Plastics	www.unimarkplastics.com
Fiskars Corp.	www.fiskars.fi
Wendland Conservatories Ltd.	www.wendland.uk.com

7. Medical, emerging and other end use applications

NDH Medical, Inc.	www.ndhmedical.com
Storopack Hans Reichenecker GmbH	www.storopack.com
Puratech GmbH	www.puratech.ch
ERBE Elektromedizin GmbH	www.erbe-med.de
Tecan Group	www.tecan.com
RTP Co.	www.rtpcompany.com
Advanced Ceramics Research, Inc.	www.acrtucson.com
Sarnatech BNL Ltd.	www.sarnatech-bnl.com
BREE Collection GmbH & Co. KG	www.bree.de
Malden Mills Industry, Inc.	www.polartec.com
Euro Projects (LTTC) Ltd.	www.europrojects.co.uk
Dow Chemical Co.	www.dowcontinuum.com
Sekisui Chemical Co.	www.sekisui.co.jp
Northrop Grumman Integrated Systems	www.northropgrumman.com
Jennmar Corp.	www.jennmar.com
Carpet America Recovery Effort (CARE)	www.carpetrecovery.org
DuPont Engineering Polymers, ATC	www.plastics.dupont.com
PlasticsEurope	www.apme.org
Delta Plastics	www.deltapl.com

Chapter 2
Plastics End Use Application Fundamentals

Keywords Polypropylene • Polyethylene • Polystyrene • Polyvinyl chloride • Polyethylene terephthalate • Markets • Applications • End use • Extrusion • Injection molding

2.1 Introduction

2.1.1 Markets Overview

Plastics play an indispensable role in a wide variety of markets, ranging from packaging and building & construction to transportation, consumer and institutional products, furniture and furnishings, electrical and electronic components, medical and others. This enormous variety of end uses for plastics materials is what gives the industry its remarkably dynamic character. This also offers an exciting and significant challenge to companies, to find the optimum mix of markets and customers to pursue that which fits with the products it makes, the services it offers, and its financial goals.

Some companies concentrate their efforts principally on the larger end use markets, but many others find it safer to spread their business over a number of application areas. The type of products a company makes defines its marketing efforts. Packaging is the largest end use, followed by building/construction, automotive, and electrical/electronic. Extrusion is the primary process used in plastics since most packaging and building/construction products sold in are extruded. Injection molding, the next most important process, is used for most complex parts. Blow molding is also extensively used in the manufacture of packaging products. 'Other' processes used are mainly thermoset processes, such as compression and transfer molding, and reaction injection molding.

While somewhat arbitrary and not always consistent within the industry, an application is typically categorized by its main attributes; therefore, an electrical

D.V. Rosato, *Plastics End Use Applications*, SpringerBriefs in Materials,
DOI 10.1007/978-1-4614-0245-9_2, © Springer Science+Business Media, LLC 2011

connector in a car wiring harness is usually considered to be an automotive application, while electrical connectors employed in a variety of end uses are considered to be electrical/electronic applications. The following segments are primary market groupings used by most industry analysts.

2.1.2 Packaging Market

Packaging constitutes the single largest end use for plastic materials, accounting for approximately 30% of all plastics used. Plastics are enjoying substantial growth as a packaging medium, at the expense of traditional materials such as glass, metals, and paper, in spite of environmental pressures on plastics materials in general, because plastic packages are lighter weight, more flexible, and easier to process. Plastic packaging also displays product more attractively than do most competing materials, offers better protection against spoilage/breakage, and provides savings in freight costs (primarily fuel economy). Plastic resins most often used in packaging are polypropylene, polyethylene, polystyrene, polyvinyl chloride, and polyethylene terephthalate, though nylon 6 is also used, largely as film in food packaging for meat. Post-consumer recycling is also an important consideration for PE- and PET-based packaging; some major end users and some government entities even specify recycled content. Because economics are frequently a principal factor in choosing one material over another, commodity-type materials tend to be favored for their lower cost. The growth of plastics packaging is due in no small part to the technologies available to convert materials into packs of different forms, flexible, semi-flexible and rigid, at high speed, which has had a crucial role in minimizing the cost of plastics packaging.

Plastic packaging can be grouped into four areas as:

- Primary packaging for the final product, in the form of bags, pouches, bottles, or other containers.
- Secondary packaging for the primary packaged product, in the form of shrink or stretch films, bottle crates, and transit containers.
- Retail packaging in supermarkets and other outlets, in the form of bags on the reel, wicketed bags, check out bags, and shoppers.
- Consumer packaging, in the form of freezer bags and cling films.

Film and containers for consumer goods and food constitute the bulk of the packaging market and are truly commodities, using commodity polymers. Polymethylpentene-I has found some specialty food-packaging applications where its unusual combination of transparency, gas permeability, and heat and chemical resistance properties can sometimes offer greater value than the commodity polymers. One of the attractive aspects of the packaging market is its relative resistance to economic cyclicality, at least in food packaging. The ongoing transition in beverage and food packaging from other materials to plastics has benefited processors, particularly blow molders, serving the aseptic and hot-filled packaging markets where global demand has more than doubled in the last 3 years. In the U.S., the area of highest growth in consumer packaging (20–30% AAGR) is polypropylene retort pouches, which are needed to keep food fresh for the military. Retort packaging is being used in brand-new applications, such as rice, tuna fish, yogurt, puddings, and even pet foods. The market is quickly moving from standard polyethylene pouches to PP retort pouches.

A number of specialty uses, such as industrial machinery and custom packaging, are much smaller in volume but offer better earnings potential to both the supplier and the user. These uses can range from made-to-order polystyrene foam protective moldings to injection molded acrylic cases for small tools. Three key packaging markets – healthcare, high-visibility, and protective – are expected to see strong growth of 4.7 to 5.2% from 2010 through 2015, with government regulations and consumer buying trends driving the increases.

With emphasis on healthcare cost control, plastics find increased use in applications in healthcare packaging. Factors influencing the increased penetration of plastics in healthcare packaging are increased use of disposable products, an aging population, continuing cost pressures on suppliers, increasing influence of hospital and healthcare-related purchasing groups, continued shift to outside contract packaging, and more and more emphasis on child-resistant/senior-friendly and tamper-evident packaging. The major plastic healthcare packaging products are segmented into two groups: medical and pharmaceutical. The former include intravenous (IV) bags, other bags and parts, kits, tubing and containers, syringes, trays, and a miscellaneous group. Pharmaceutical packaging products are made up of closures, bottles/vials, blister packaging, and a miscellaneous category. Bottles/vials, tubing/containers, syringes, kits, are estimated to be the leading healthcare packaging applications, accounting for over 83% of total plastic volume in 2010.

In pharmaceuticals, blister packs will grow fastest followed by plastic bottles. Unit-dose regulations for institutional drugs, along with clinical trials and over-the-counter medication, are factors influencing blister packs' strong 6.3%/year growth.

In medical packaging, several major producers of high-tech, highly layered flexible films for barrier food wrap have recently qualified their first coextruded films for medical packaging. Food-film companies had tried for years to apply their low-cost production, economies of scale, and experience with highly layered structures to the medical market. Now they are becoming factors to reckon with in this high-priced specialty business. Requirements for medical packaging are very different from those for food wraps. Flexible food packages have to provide tight oxygen barrier to preserve processed meats and cheese. Extruded films for medical packages have to provide a microbial barrier, resist puncture, and survive sterilization. What the food film companies have to offer is expertise in highly layered structures. Coextrusion cuts cost in several ways. It can replace expensive materials and use resins more efficiently. It can also in some cases reduce the number of production steps by eliminating coating or laminating.

In high-visibility packaging such as clamshells and blisters, consumer spending on larger and costlier items, coupled with retailers' needs to deter theft while letting buyers inspect package contents will drive growth of 5.1%/year. Demand for clamshells will remain strong, despite their high cost, for the benefits they provide in packaging large, heavy products as well as multiple components. They can also be displayed on pegs as well as shelves.

Burgeoning internet and mail-order sales along with specialized packaging needs in electronics, medical, and other markets will increase demand for protective packaging by 5.2%/year. But there will be gains for foamed rolls, and inflatable bags will post the fastest growth. Foams will grow 5.3%/year, while protective mailers will gain 5.0%/year, insulated shipping containers 6.2%/year, and inflatable bags 8.1%/year.

The digital video disc (DVD) market has also inspired packagers' attention. The DVD has been a breakaway winner in the multibillion-dollar home entertainment industry. More than 60% of U.S. households now own a DVD player, and by 2011 so will more than 67% of European homes. Money spent buying and renting DVDs overtook spending on VHS cassettes in 2010, and by 2015 will account for 93% of spending on video software. DVDs are purchased at much higher rates, making packaging more important to consumers. DVD packaging is almost exclusively produced in polypropylene.

2.2   Other Major Markets

2.2.1   Building and Construction Market

Building and construction is the second largest end use market for plastics with 28% of plastics materials consumed in this market sector. As in packaging, plastics have displaced such traditional materials as wood, glass, and metal, based on improved performance and lower cost. For pipes, valves, and fittings, plastics offer superior corrosion resistance and are lighter, easier to install, and more cost-effective than their alternatives. Impervious to chemicals and sulfur-bearing compounds, plastic piping safely transports everything from fresh water to salt water, and from crude oil to laboratory waste. These qualities also have combined with plastics' high strength-to-weight ratio to produce materials for bridge and other infrastructure construction, including tough reinforcement rods, nonskid surfacing and quickly installed replacement decking. In this distinctly cyclical industry, most of the volume usage is processed by extrusion, such as for piping, conduit, wire insulation, siding, reservoir liners, erosion control netting, or architectural sheeting. Sometimes this category includes such agricultural end uses as irrigation pipe and fittings, mulch films, and fencing. For market participants, there are a number of regulatory hurdles to overcome such as the Underwriters Laboratories (UL) listings and the National Sanitation Foundation (NSF) listings for both potable and waste water. For proprietary processors, parts must comply with applicable building codes at the local, regional, and national levels.

In the construction of buildings, plastics abound in plumbing fixtures, siding, flooring, insulation, panels, doors, windows, glazing, bathroom units, gratings, railings, and a growing list of both structural and interior or decorative uses. For commercial buildings that contain sensitive electronic equipment, plastics can provide a highly protective housing that does not interfere with radio frequency or magnetic waves. In residential buildings, single-families increasingly opt to refurbish existing older homes to suit their growing needs. Remodeling contractors and do-it-yourselfers increasingly turn to light, strong, easily handled plastics to make improvements and repairs, from ceiling tiles to advanced wiring and modern bathroom fixtures. In building additions as well as new construction, bathroom fixtures such as tubs, showers, and sinks can be constructed in one piece, walls, pipes, and all, then positioned and attached to the building frame, producing significant savings in construction and installation costs. In older home restoration, plastics can provide the best of both worlds duplicating yesterday's beauty using today's superior materials. Architectural touches such as decorative wall, door and window moldings can be achieved that mimic marble or hand carvings but offer the ease of care and damage resistance.

2.2.2 Automotive Market

Durable, lightweight and corrosion resistant, plastics offer fuel savings, design flexibility and high performance at lower costs to designers facing today's complex transportation needs whether on land or sea, in the air or in space.

Auto makers choose plastic parts for their durability, corrosion resistance, toughness, ease of coloring and finishing, resiliency and light weight. Plastics have significantly reduced the weight of the average passenger car saving millions of gallons of gas each year. Plastics have found their way into automobile components such as bumpers, fenders, doors, safety and rear-quarter windows, headlight and sideview mirror housings, trunk lids, hoods, grilles and wheel covers. Automobile designers also discovered that plastics solve one of their most complicated design problems, namely, what to do with the fuel tank. Using plastic gives them the freedom to fit tanks into the overall concept rather than designing around the unwieldy but essential part. Advances in engine technology, fuel management, and emissions requirements continue to drive up under-the-hood temperatures. Where certain high-temperature plastics were adequate a few years ago, today that is no longer the case. These automotive advances are driving the use of higher end plastics under the hood, as well as the development of new plastics and coating to better perform in new under-the-hood environments. In automotive interiors, plastics used in flooring, seats, dashboards, and paneling maintain their attractive appearance and are easy to clean.

Plastics versatility is aiding the automotive industry to meet ever more challenging requirements in terms of economical performance, safety, comfort, and environmental considerations. The automotive industry is the single largest end user for many engineering plastics, such as nylon, polycarbonate, acetal, or modified polyphenylene ether. It is also an important market for commodity polymers such as PP, PE, and PVC. Automotive business is notoriously cyclical, and suppliers can easily find their orders canceled literally overnight if demand takes a downturn. Automotive business cycles not only include the ups and downs of the overall economy but also those of individual brands and models. Operating successfully in a cyclical industry requires companies to have considerable flexibility with respect to manufacturing capacity, and financial reserves. Nevertheless, companies have prospered by learning how to cope with these problems successfully. Recent industry estimates expect the use of engineering thermoplastics in exterior automotive applications to grow by 4.8% annually through 2010.

2.2.3   Electrical and Electronic Market

The electrical and electronic (E/E) end use market is the third largest market for plastics applications ranging from miniature connectors to large housings. An ever-growing universe of electronic equipment, components, and gadgets is expanding the world and improving lives. In this age of electronics, computers power the business world and teach skills to toddlers. Communications systems reach the far corners of the earth and beyond, tasks that once took many hours now can be accomplished in minutes and leisure hours have become more varied. In E/E uses in the home, plastics with premium thermal and insulating properties are used to insulate nearly all house wiring and are also used in electric switches, connectors, and receptacles. Lightweight, durable, attractive, and cost-effective plastics are used in nearly all small appliances from coffee makers to hair dryers and shavers. The materials are also indispensable in major appliances, which would cost at least 25% more and use 30% more energy than similar products produced without plastics. All refrigerators today, for example are insulated with thermally efficient plastic foam, and their interiors are made of durable, easy-to-clean plastics.

An important aspect of the electrical/electronic market in many applications is the fast time-to-market demands of the industry of 6 months or less and the relatively short product life cycles from one to one and a half years. This requires that suppliers work with end users on new applications from the beginning, but this also prevents other suppliers from displacing or even sharing business with an existing supplier except in the case of major quality or delivery failures. Cell phones are a typical example of such applications. Without plastics, most of the electronic products used today would not be practical or economical. Designers of computers and business equipment choose plastics for their toughness, durability, ease of fabrication into complex shapes, and their electrical insulation qualities. Plastics have been fundamental to electronic progress for decades, housing electronics, insulating components from all types of interference, and protecting parts. In fact, microprocessor miniaturization would not have been impossible without the quality and cost-effectiveness of plastics. The continuing miniaturization of circuit boards and components such as computer chips increasingly relies on high-performance plastics to provide tough, dimensionally stable parts that can withstand both the stress of assembly and the strain of use. With plastics, electronic designers simultaneously can decrease size and increase the functionality of circuitry in consumer, business, and industrial electronics. In the modern factory, automated production relies on plastics for control panels, housings, printed wiring boards, sensors, and robotic components. Corrosion-resistant, flexible plastic also serves as conduit for electrical wiring.

The E/E market requires that both materials and parts for each new product be evaluated against Underwriters Laboratories (UL) requirements. Materials must be offered in flame-resistant formulations if they are not already inherently flame resistant. Regulatory requirements in Europe also mandate that flame resistance be achieved without the use of halogenated components. While a UL listing for flame resistance for a qualified product is relatively quick and inexpensive to get, another important UL requirement, the limiting temperature index (LTI), which specifies the maximum continuous use (operating) temperature for a material typically takes from one to one and a half years to obtain. The expense of obtaining and maintaining these listings demands that the market potential for each product be sufficient to justify them.

2.3   Key Crossover Markets

2.3.1   Consumer Goods Market

Consumer products cover many applications, from cosmetics and toothbrushes to housewares, toys, recreational goods, furniture, lawn/garden equipment, and others. Considering the diversification of this category, it is extremely difficult to assign overall growth figures though there are a number of very profitable niche opportunities. Trends in disposable personal income play a major role in the purchase of these products with any growth in real wages boosting consumer spending in this sector. The downside of business in consumer goods is that the product life cycles are often short, and most applications are price sensitive. Also, many of the end uses have seasonal patterns. Consumer non-durables are defined as items expected to last less than 3 years and consist mostly of single use, disposable items, such as single use cups, picnic tableware, and disposable diaper components, and specialty packaging (e.g., cosmetics cases). Cost is king, as one would expect in such commodity applications.

Although synthetic fibers are not widely thought of as part of the plastics industry, they are and strongly affect the supply and prices of their base resins. The great majority of fibers are used for consumer goods, such as carpeting and clothing. When housing starts or consumer purchasing passes through a dip in their economic cycles, integrated polymer/fiber producers have more capacity to divert to plastic resin sales.

2.3.2   Industrial and Stock Shapes Markets

Industrial components mainly deal with machinery parts, such as pump housings and impellers, conveyor links, or gears and bearings. This market is highly fragmented with low competitive visibility and high value in use, and therefore high profitability. Product life cycles are typically long. End users are often local, and volumes often tend to be small. The materials used are mainly engineering and high-performance polymers though the full range of plastics from commodities to specialties are employed in this sector.

Rod, tubing, and sheet are typical semi-finished shapes which are mechanically fabricated into other parts for a range of unidentified end uses. Semi-finished shapes are often used to make prototype parts for evaluation or even small numbers of commercial parts. Though this market segment is important in terms of size, its growth rate is less than many others and it is very competitive with considerable consolidation taking place in recent years. Shape producers rarely have direct contact with end users, their route to market being almost entirely through stock shape distributors, a group separate from plastic material distributors.

2.3.3   Other Markets

Several other major market segments include (1) medical and (2) aerospace and military.

The medical market is a fast growth, recession proof plastics market application area. Medical products typically are divided into two categories, disposables and durable equipment. Though medical disposables are commodity-type products with emphasis on cost, reliability demands are high making profit margins much better than most commodity markets. Many medical disposables such as catheters, blood bags, IV bags, etc., are relatively high tech. The high tech nature of durable medical equipment also provides the opportunity for higher margin business, but as reliability demands are also high, exposure to product liability claims can be significant.

Aerospace and military end use areas are among the most challenging plastics markets. Market end users are highly fragmented, volumes are relatively small to medium, and application developments have very long timelines, though product life cycles tend to also be long. Once a material is approved, a supplier can usually look forward to a very long run of many years to even several decades. Competitors are unlikely to be able to qualify their material as an alternate source. Parts contracts however are generally put up for bids annually, so that a processor must defend the business every year.

Chapter 3
Electrical and Electronic End Use Applications

Keywords Electrical • Electronic • Fuel cell • Transistors • EMI • LCP • PEEK • Office products • Appliance • Recycle

3.1 Electrical and Electronic Applications

3.1.1 Electrical and Electronic Applications Introduction

Plastics is an indispensable material for the electrical and electronics sector (E/E). Many of today's new technical developments capitalize on the latest versions of new generation plastics. As a result, devices such as cell phones are becoming smaller and lighter. Lightweight, durable, attractive, and cost-effective plastics are also being used in nearly all our small appliances. Some sectors of the E/E market have been evolving toward becoming commodity businesses but there are still a number of high-performance, attractive opportunities in specialty sectors. A need exists for higher temperature resistance materials with processing characteristics suitable for use in thinner walled, smaller components. Even the largest subcategory of the E/E market, wire and cable, which has a range of commodity, and semi-commodity applications still has many specialty applications, and new product and application development across the E/E sector continues to result in significant, profitable business as performance requirements keep ratcheting up.

3.1.2 Fuel Cell Plates

Equipment and materials suppliers are making progress in the compression molding of thermoset bipolar plates for fuel cells. Bulk Molding Compounds Inc. (BMCI) says graphite-filled thermoset vinyl ester bipolar plate technology offers the best overall performance, productivity, and cost for use in PEM (proton exchange

D.V. Rosato, *Plastics End Use Applications*, SpringerBriefs in Materials,
DOI 10.1007/978-1-4614-0245-9_3, © Springer Science+Business Media, LLC 2011

membrane) fuel cell applications compared to metallic and other composites. Conductivity of the plates is now around 70 Siemens/cm versus 20–30 S/cm for plates launched a few years back. BMCI attributes this to more homogeneous distribution of conductive additives in the compound. In addition, BMCI has introduced a new high-temperature cure (370°F) package, which reduces cure times by 50% (to around 15 s) and eliminates the 15–20 min post-bake required to remove residual volatiles such as styrene monomer.

BMCI has also developed a vinyl ester adhesive/sealant for use in bonding the cooling channel side of the plates from adjoining cells to form a well-sealed path for the stack coolant. Like the bipolar plate compound, this adhesive is also highly conductive. Compared to other adhesives used in this application, it is also substantially less expensive and reduces the voltage drop between plates by 30–50%. As fuel cell usage becomes commercial and volumes grow, the recycle of fuel cell components will become an issue. BMCI has completed regrind trials with very encouraging results and work is currently underway with companies such as Plug Power, Inc. and General Motors.

3.1.3 PLED Flat Panel Displays

Cambridge Display Technology (CDT), a global leader in the research, development, and commercialization of light emitting polymers, targeted for use in electronic display products used for information management, communications, and entertainment, is working jointly with Sumitomo Chemical (www.sc-sumitomo-chem.co.jp) to develop/scale-up a new range of solution processable polymer light emitting diode (PLED) materials designed for use in flat panel displays. Features include reduced power consumption, size, thickness, and weight, very wide viewing angle, superior video imaging performance, and the potential for use on flexible display substrates. The companies are focusing on new solution processable, phosphorescent materials, such as dendrimers, which exhibit very high efficiencies and good stability. Dendrimers are nearly perfect monodisperse macromolecules with well defined, starlike structures. High-efficiency emitters are especially important in printable portable electronic displays for devices such as cell phones and digital cameras, where battery supplies are used. The materials are also expected to find applications for TV displays and lighting devices.

In the US, companies operating in the same field as CDT include Gyricon, a spin-off of Xerox, and E-Ink. Their displays differ from CDT's in that they use reflection rather than transmission, and are easier to apply onto flexible substrates. The displays entail a layer of microscopic spheres of black and white particles with negative and positive charges respectively sandwiched between two films. Depending on the charge put across the films, the spheres look black or white and can make a monochrome image. E-Ink has already demonstrated color images in a similar way. Gyricon is targeting remotely updatable signage, and E-Ink is concentrating on portable e-readers. Its technology is currently used in display for Sony's Librié e-book reader. CDT is mainly concerned with displays in more conventional equipment like laptops.

3.1.4   Flexible Printed Transistors

Plastic Logic Ltd., a leading developer of plastic electronics including printed flexible thin film transistor (TFT) arrays, is developing and exploiting a portfolio of intellectual property based on inkjet printing of active electronic circuits using advanced plastic materials to form thin film transistors that can be used in active matrix back planes that drive displays. Other potential applications include smart labels, smart packaging, and radio frequency identification (RFID) devices. Plastics electronics can be produced directly at high speed from CAD data to large, flexible surfaces using ink jet printing equipment rather than complex photolithography and vacuum systems used to make today's transistors. With the process' low temperatures, substrates used can also be plastic.

The company recently announced a non-exclusive agreement with E Ink of Cambridge, MA, the leading developer and marketer of electronic paper display technology, to cooperate on the design and fabrication of flexible all plastic electronic displays. E Ink and Plastic Logic will to combine their technologies to produce high-resolution active-matrix displays suitable for applications from smart cards and cell phones to wireless electronic readers (e.g., e-newspapers, e-books, e-maps). The liquid crystal active-matrix displays commercially available today are produced using two sheets of heavy, fragile glass and require multiple vacuum deposition and high precision lithographic steps. In contrast, the E Ink and Plastic Logic displays are flexible, thin, light weight, bright, high contrast, shatterproof and can be produced at low cost in high volume. Plastic Logic is currently expanding its existing plastic electronics mini-lab, installing a prototype line, and is targeting for further production scale up.

E Ink provides electronic ink in sheets of imaging film that are ideally suited for flexible display applications due to their thin form factor and inherent flexibility. E Ink's electronic ink is an image stable reflective display technology that uses ultra-low power but is easily read under any lighting condition including direct sunlight. Unlike liquid crystal displays, the image on E Ink displays looks the same from all viewing angles and will not distort when touched or flexed, making them suitable for flexible displays and portable devices. E Ink Imaging Film entered mass production last year as an enabling component in the unique SONY LIBRIé electronic reading device currently available in Japan.

3.1.5   High-Performance Ignition Coil Bobbins

MSD Ignition, a world leader in performance auto ignition technology, is moving up to DuPont Thermx PCT high-performance polyester for coil forms in its ignition coils. The first MSD coil to make use of Thermx is its Blaster SS, found in both automobiles and NASCAR racers. With efficient E-core windings, the Blaster SS generates 300 mA at up to 40k V. The coils, with a bobbin molded from PCT, have withstood tests at 70,000 V, the limit of MSD Ignition's test rig, with no arcing, or cracking of the coil form.

MSD Ignition is transitioning to Thermx PCT for bobbins in all high-performance racing/street use coils. In new designs, PCT also allows smaller bobbins and reduced coil size. The material has excellent dielectric properties at elevated temperature and a high melting temperature. Vibration resistance is another key benefit of the PCT coils. Most MSD ignition coils are potted in epoxy, which adheres well to the PCT. The result is less vibration separation. The Blaster SS coil's housing and cover are both molded from DuPont Rynite FR530 PET polyester, which provides good strength, stiffness, and toughness at under the hood temperatures. MSD Ignition has adopted PET for a range of coil forms, housings, and other ignition system components. At the Atlanta Motor Speedway NASCAR series, the three lead teams of Dale Earnhardt Jr., Jeremy Mayfield, and Kasey Kahne all used MSD 6HVC Ignition Control.

3.1.6   Smart Wallplates

Cooper Wiring Devices subsidiary of Cooper Industries, a worldwide manufacturer of electrical products and tools/hardware, has switched to Bayer Polymers' Makrolon PC for use in its new line of mid-size thermoplastic wallplates. Polycarbonate was selected, versus other materials such as thermoset urea, for its processing ease, high structural rigidity, and superior resistance to warping, impact, and abrasion. Using this material in an injection molding process allowed Cooper to reduce cycle times and eliminate costly secondary operations required to remove flash during the molding process. The company's new line of mid-size wallplates adds a full 0.375″ on all sides to NEMA standard-size plates, providing needed gap coverage for over cuts or other large wallboard-to-box openings. With polycarbonate, Coopers' new line of wallplates has a high gloss finish and is resistant to soil adherence. In addition, the wallplates are easily cleaned in comparison to other material surfaces. The line is UL listed and CSA certified, and meets NEMA WD-1 and WD-6 compliance. The thermoplastic wallplates are available in multiple colors and various configurations for toggle switches, receptacles, decorator, blank, power receptacles, telephone/coaxial cable, combination devices, and pre-marked special uses.

3.1.7   Electronic Cloth

Researchers at the University of California at Berkeley in an industry first have printed plastic pentacene transistors directly onto cloth fibers. The approach may be used to embed sensing, actuation, and displays into clothing and surface coverings.

The process, which is compatible with textile manufacturing, does away with expensive masks, or negatives, which are normally used in printing circuits on silicon. Instead, the researchers developed a new mask that is made up entirely of textile fibers. The transistors created by the Berkeley research team were fabricated using a 125 µm diameter aluminum wire as the gate line. This wire can be woven directly into an 'e-textile' to serve as a gate interconnect. The fiber was masked with orthogonal overwoven 50 µm diameter wires, which served as channel masks with a transistor formed at every intersection, while 100 nm gold was evaporated to form source or drain contacts. All the transistors were ~125/50 µm, since transistors were formed only on the surface of the gate wire, which was exposed during pentacene evaporation. The transistors that resulted are similar to conventional inverted top contacted pentacene thin film transistors (TFTs). However, the entire transistor was formed on a metallic gate wire, permitting easy e-textile integration to create 'smart cloth'.

3.1.8   *Intrinsic EMI Shielded Products*

Autotronic Controls Corp.'s (ACC) new MSD digital ignition tester used to check ignition performance in racing cars and other high powered vehicles has a housing molded from Dupont's Zytel EMX nylon shielding resin. The material effectively shields sensitive microprocessor circuitry against electromagnetic interference (EMI), or noise. ACC evaluated an unshielded housing but found that electrical noise caused problems with the unit. The Zytel EMX provides effective shielding in a sturdy housing produced using standard injection molding technology at a lower total cost than using coated shielding. The new housing show virtually no noise problems during field trials, a significant accomplishment since MSD ignition coils used in high powered racing cars can generate up to 50,000 V of electrical potential in repeated bursts, generating intense electrical noise.

Racing crews use the MSD ignition tester to check ignition control and coil operation without having to remove them from the car. The tester produces a simulated trigger signal that fires the car's ignition coil just as if the engine were running. If the spark fails to jump the gap in the tester, there is an ignition problem, which the tester can help trace down. With the built-in housing shielding protection provided by Zytel EMX, Autotronic Controls eliminates the cost of applying a conductive shielding coating required with a conventional plastic housing. While a metal housing would have provided effective shielding, it would have required a greater tooling investment and costly finishing steps for each part.

3.1.9   *Miniaturized DSL Transformers*

Epcos AG, a major developer and manufacturer of electronic components, chose DuPont Zenite LCP for its coil formers to produce SMD transformers used in telecommunications equipment. This liquid crystal polymer has the high dielectric strength/thermal resistance, and good melt flow properties essential for making miniaturized electronic components for mounting on circuit boards using SMD technology. Digital subscriber line (DSL) technology is one of the fastest growing technologies for broadband internet access and calls for interface transformers with minimal signal distortion and loss. Epcos' specially developed EPX and EPO series core geometries made of T66 and T57 ferrite materials offer good data transmission rates combined with greater miniaturization.

Dupont's Zenite 7130 glass fiber reinforced LCP combines good toughness with a heat deflection temperature of 289°C and high dimensional stability to make the coil formers for these applications. As the material uses no halogen additives, it is recognized by Underwriters Laboratories as UL 94 V-O at 0.75 mm. Since the injection molded parts do not need to be deflashed, processing costs with LCP is less than that with thermoset resins. The company can injection mold very fine coil formers with a 0.4 mm wall thickness and a 25 mm^2 footprint. The liquid crystal polymer's high strength and impact resistance allows these precision parts to withstand the stresses that occur during coil winding and the material's high thermal deflection temperature allows parts to withstand temperature peaks up to 260°C, which occur during reflow soldering, when the transformers are mounted on the circuit board.

3.1.10   Electrodynamic Loudspeakers

Cabasse, an ultra high yield speaker systems manufacturer based in France, has turned to Victrex PEEK polymer to produce the spiders on its latest range of professional loudspeakers. This innovative use of the high-performance polymer allows Cabasse to significantly improve the power capacity of the transducer. An electrodynamic loudspeaker is a transducer with a very poor yield. All parts surrounding the movable unit inside these 1,000 W speakers (acoustic diaphragm, coil, spider and front suspension) must exhibit outstanding heat resistance, as 97% of the electrical power delivered by the amplifier is dissipated in the form of heat. Spiders are traditionally manufactured from a cotton fabric coated with phenolic resin. However, at full power, Cabasse has measured temperatures as high as 270°C in the coil area of its new range of professional speakers. With a very high melt temperature (343°C), and a continuous operating temperature of up to 260°C (UL746B), monofilament yarn made of Victrex PEEK polymer gave the company the necessary high-temperature resistance.

An additional factor Cabasse's in material selection was strength and fatigue resistance, as guide parts such as the spider must withstand significant displacements. The PEEK polymer fabric delivers an ideal combination of stiffness, elongation at yield and very high fatigue resistance and the polymer's inherent ease of processing has also simplified Cabasse's manufacturing process and improved productivity.

3.1.11   Visual/Tactile Electronic Enclosures

EXO, an overmolding system developed by Inclosia Solutions, a unit of Dow Chemical Co., is an innovative technology that allows fur, leather, various fabrics, and even wood and metal to be incorporated into the surface of casings for portable devices such as cell phones, laptop computers, and pocket calculators and brings new opportunities for co-branding to original equipment manufacturers (OEMs). In a market where state-of-the-art electronics performance features are no longer the only differentiating factors in electronics purchase decisions, branding and appearance enhanced by molded in fabrics and other materials are increasingly important drivers in motivating consumer purchase.

The family of EXO system solutions developed by inclosia is designed for high volume production of a wide range of enclosures, a variant of two shot overmolding allows virtually any type of material to be incorporated in a covering while meeting cost, quality, durability, and volume requirements critical for production of electronic devices. The process starts with placing a pre-cut decorative insert into an injection mold. Once molding starts, the first shot covers the insert with a thermoplastic substrate, which also forms the enclosures structure as required by the application. An optional second shot can be used to encapsulate the edges of the back molded material. The overmolded material is permanently bonded on the outside of the enclosure so there are no issues with frayed edges compromising the integrity of the internal components. As one mold can be used for different materials/fabrics, retooling is not necessary between covering changes, which further increases speed and cost advantages.

3.1.12   High-Performance Outdoor Cable

Sporting events and rock concerts are frequently electronic extravaganzas requiring enormous quantities of cable for lighting, sound, and other media. Now, Flexalloy, an advanced PVC-based TPE from the vinyl division of Teknor Apex, is allowing the Entertainment Division of Coast Wire and Plastics Technology, a cable manufacturer specializing in these applications, to supply cable that is lighter, more flexible, and more resistant to extreme cold than cable produced using conventional compounds.

Flexalloy-based cable products have half the weight of other cable products making it possible for users to dramatically increase overhead lighting without exceeding the limit to the weight that an event ceiling structure can support. Based on ultra high molecular weight (UHMW) PVC resin, Flexalloy compound also provides greater flexibility than conventional vinyl, even at low temperature, a critical advantage at outdoor winter events such as the winter olympics. Even in warmer environments, flexibility is important for timely wiring to be used at crowded entertainment venues. For these applications, Coast Wire has introduced a new line of entertainment cable 'FlexOLite Touring Cable' based on Flexalloy compound. Flexalloy products provide the elasticity of non-vinyl TPEs, plus greater toughness, abrasion resistance, and low-temperature performance than conventional vinyl.

3.2   Office Products Applications

3.2.1   Office Products Applications Introduction

As the economy continues to rebound, tech spending for computers and peripherals has increased as well, although not to pre-2008 levels. The computers and peripherals segment showed a 14% annual increase from 2005 to 2008. PC shipments globally grew by 13.5% in 2008, bringing the market size to more than 175 million units. Growth rates for 2010 and subsequent years are forecast to rise more slowly, with shipments expected to grow by nearly 11% in 2010 and about 8% annually from 2011 to 2015. Wireless capabilities, falling prices, and the growing need for mobility have increased demand for notebooks; however, many buyers remain price sensitive and continue to evaluate the benefits of mobile computing. While portable PCs continue to grow much faster than desktops, recent results demonstrate the continuing competitiveness of desktop systems. Worldwide shipments of desktop PCs grew by 13.4% year-on-year in 2010, up from 9.6% in 2009, while growth in portable PCs slowed to 28.5% in 2010, from more than 35% in 2008. Desktops continue to represent more than 70% of worldwide PC shipments, and their strong growth helped boost growth expectations for total PC shipments in 2010.

3.2.2   Miniaturized Fuel Cells

Trends in portable electronics require longer lasting power, more energy than current technologies might be able to provide. MTI MicroFuel Cells Inc. is developing miniaturized fuel cells suitable for handheld electronic devices to replace power packs currently used by OEMs in many rechargeable handheld electronic devices like PDAs and smartphones. Mobion cord free rechargeable power packs are based on direct methanol fuel cell (DMFC) technology. A recent Allied Business Intelligence (ABI) study sees almost 22% of all handheld electronic devices with rechargeable power packs being powered by DMFCs by 2011.

MTI Micro's DMFC system, based on work contained in more than 50 patents, consists of a unique fuel cell array, custom fuel feed control and user replaceable methanol fuel cartridges, is intended initially for integration into industrial handheld electronic OEM devices. Product launch is planned for early 2011. Under a strategic alliance agreement, MTI Micro has incorporated DuPont's membrane electrode assembly in the core of the proprietary fuel cell system. The best-known and most widely used membrane material used in this application today is Nafion, a polyperfluorosulfonic acid product that is cast into film and supplied by DuPont Fluoroproducts. Flextronics will assemble this initial product at Flextronics' specialized 'Product Introduction Center' in San Jose, CA. The power pack technology is designed to increase run time (time between recharges) two to ten times over that of existing battery technologies.

At the core of Mobion technology is its unique approach to managing water that is produced by the chemical reaction at the cathode, and required for the chemical reaction at the anode. MTI Micro's Mobion technology simplifies the traditional approach in which water must be externally pumped from the cathode to the anode side for the chemical reaction to occur and energy to be produced. Using MTI Micro's proprietary technology the water required for the process on the fuel side is transferred internally within the fuel cell from the site of water generation on the air side of the cell. This internal flow of water takes place without the need for any pumps, complicated re-circulation loops or other micro-plumbing tools. As a result of the water reallocation internal to the fuel cell, the system can accept 100% 'neat' methanol and the elimination of 'micro plumbing' makes more room available within the pack for the fuel. The net result is that Mobion fuel cells are easier to manufacture. They are also more compact, making Mobion-based DMFCs small and light enough for handheld devices.

3.2.3   Rollable Reading Displays

Lightweight, unbreakable, large area displays that can be rolled up into a small housing when not in use are particularly attractive for mobile applications. Work is progressing at PolymerVision, a Philips Technology Incubator that will transform the way information is viewed, particularly when we are on the move by providing flexible displays not much smaller than a computer, that roll up and tuck away when not in use. After years of work on polymer electronics Philips is producing prototypes of ultra-thin, large area, rollable displays and intends to move rapidly toward a commercial production process. One prototype, designed for e-newspapers, suggests a dual screen foldable stick mechanism to allow dual functionality of the product as both e-newspaper and e-book. News can be downloaded via subscriptions direct to a PC inbox, or uploaded from news kiosks as one-time purchases. Easy one-button browsing and simple elegant design would appeal to both traditional reader and early technology adopters. The displays combine active matrix polymer driving electronics with a reflective 'electronic ink' front plane on an extremely thin sheet of plastic. Ultimately, flexible large area displays that can be integrated into everyday objects like a pen could be feasible. Such displays would greatly stimulate the advance of electronic books, newspapers, magazines, and news services; applications that currently depend on fragile, large laptops or small, low-resolution displays of mobile phones.

PolymerVision has made organics-based QVGA (320×240 pixels) active matrix displays with a 5 in. diagonal, a resolution of 85 dpi, and a bending radius of 2 cm. The displays combine an active-matrix back plane containing polymer driving electronics with a reflective front plane of 'electronic ink' developed by E Ink Corp. Electronic ink based displays can have thin flexible construction, as they do not require cell gap control. Displays made with electronic ink technology are ideal candidates for reading intensive applications because of their excellent, paper-like readability and extremely low power consumption. Currently, PolymerVision is capable of producing over 5,000 fully functional rollable display samples per year and is in the process of defining a pilot production line.

3.2.4   Digital Pens

With Logitech's new io_2 Personal Digital Pen, it is possible to take notes at a meeting with pen and paper, return to the office and transfer the handwriting to computer. This application provides development opportunities for polycarbonate, PPO/PS, and soft touch TPE type engineering plastics. The Logitech io_2 retail package consists of the pen, a USB cradle and recharging station, Logitech io_2 software, special preprinted digital paper, an AC adaptor and five ink refills. The pen software is compatible with Windows 98 or higher.

An optical sensor located beneath the $200 electronic pen's ballpoint tip captures the handwritten notes by recording a pattern of dots created when the pen is used to write on special digital paper. The digital paper (with Anoto functionality) is created by printing a proprietary pattern of very small dots on paper that is perceived by the eye as slightly off white in color. The printed dot pattern locates words and images on the paper and assures they appear in the same place in digitized files. The pen, used as a regular ballpoint pen, is activated by removing the cap and deactivated by replacing the cap. When in continuous use the pen's battery has typically a 3-h life. Up to 40 pages of handwriting can be stored in the pen's 856-kB flash memory. The captured writing can then be transferred to a computer by placing the pen into a dock connected to the PC via a USB port. The handwritten images can be pasted into various applications including word. The Logitech io_2 pen is a new product category and represents a platform with an enormous potential for growth particularly as good handwriting recognition software becomes available.

3.2.5   Orbital Web Cameras

Logitech, the world leader in Web cameras, has a new web camera, Logitech QuickCam Orbit, that offers unique features including mechanical pan and tilt as well as an elegant form. The camera combines overmolded TPO to a polycarbonate shell base in its unique sleek design. It is ideal for video instant messaging, since it physically moves to keep a person's face automatically centered in the field of view. A heavy base contains the motor to swivel the camera, topped by a globe slightly smaller than a tennis ball, which houses the camera and a red LED behind a clear plastic shield. The camera is built into the globe and positioned on a motorized mount that tilts up and down.

Logitech designs, manufactures, and markets personal interface products that enable people to effectively communicate in the digital world. The Logitech QuickCam Orbit is the industry's premier webcam. The QuickCam Orbit mechanical pan and tilt feature allows the webcam to physically turn 128° side-to-side and 54° up-and-down, freeing users from manual camera maneuvering. Logitech's face-tracking software enables the camera to automatically follow a person's face, keeping it centered in the field of view. Auto-zoom technology offers the option to zoom in up to three times the normal viewing size. QuickCam Orbit captures up to 1.3 megapixel still photos, and true 640×480 videos with its high-quality CCD sensor. A 9-in. stand is included, so the camera can be elevated to eye-level, and also a built-in microphone for easy addition of audio to instant messaging or video clips.

3.2.6 Removable Data Storage Media

Imation Corp., a worldwide leader in removable data storage media, has introduced its new double layer DVD+R media, the industry's next advancement in DVD technology, to meet the increasing storage requirements of consumers for high-capacity applications such as digital video. The new disc nearly doubles the storage capacity of a DVD recordable disc from 4.7 GB to 8.5 GB on a single side, without having to turn the disc over during recording while remaining compatible with DVD video players and DVD-ROM drives.

The new media achieves the higher capacity by using a dual layer DVD+R system with two thin embedded organic dye films for data storage, which are separated by a semi-reflective layer and a spacer layer. The recording principle is based on irreversibly modifying the dye's physical and chemical structure, induced by heating with a focused laser beam. The recorded marks have different optical properties than their unmodified surroundings, giving the same read-out signals as read-only discs, based on the length of the recorded marks and unmodified spaces between them. According to technology developers Philips and Mitsubishi Kagaku Media, a key challenge to overcome was to minimize the interference between data recorded in the first recording layer (L0) while recording on the second layer (L1). In addition, the first layer requires the use of a transparent thin metal layer that should still have enough cooling power allowing controlled recording of data in the first layer. Philips in cooperation with MKM were able to reach major innovations in the materials used for the spacer layer, the transparent metal reflector, and recording layers while remaining within the boundaries set by the requirements for backwards compatibility with double layer DVD-ROM.

3.3 Appliance Applications

3.3.1 Appliance Applications Introduction

A continued strong new-home market and an equally strong remodeling market is keeping the demand for major and small appliances steady, with nearly all categories of large appliances expected to register a minor uptick in units sold. To propel the appliance market into bigger and better times, many of the major manufacturers are moving into the very high-end market. GE Appliances, Whirlpool, and others are making a push into an area of the market where price points near $2,000 are common for big-ticket durables such as refrigerators, freezers, clothes washers/dryers, and versatile stove/oven/refrigerator combos. Added features in these 'smart appliances' are creating a value that the major appliance makers hope will translate into bigger profits.

The other end of this market is crowded with products in the $500 price range, with entrants from China and Korea (LG Electronics, for one) squeezing this end of the market. Competition in the small appliance category is not much better, with small appliance manufacturing just about nonexistent in the US. Much of that market went to Mexico a decade ago, and what did not move south then has now moved to the Far East—China particularly. Mexico is still home to a number of appliance manufacturing plants, where companies such as Maytag, Bissell, Hoover, Eureka, and Whirlpool, among others, have maquiladora plants on the border.

3.3.2 E/E Design for Recycle Appliances

Japan's home appliance recycling legislation, making it mandatory for manufacturers to charge consumers for collecting old appliances, is transforming appliance material selection and design. The legislation sets minimum recycling rates by appliance weight; with 60% of an air conditioner's weight to be recycled, 50% of a refrigerator, 50% of a washing machine, and so on. It is expected that these rates will be increased to 80–90% by 2011.

In response to the legislation Japan's appliance makers are making their products more readily recyclable by using fewer plastics and grades, and plastics that are easier to recycle. PS, considered easier to recycle, is used rather than ABS over concern that toxic substances derived from acrylonitrile may be emitted if ABS is incinerated. Both Fujitsu and Matsushita switched from using ABS in air conditioner drain pans to an alloy of syndiotactic polystyrene (SPS) and polystyrene. A PS/SPS alloy is also being used in place of glass fiber reinforced PBT in a vacuum cleaner blower. Similarly, polypropylene is replacing ABS in vacuum cleaner housings. In refrigerator designs, trays now are more likely to be chemical-resistant GPPS than AS resin and PVC is also being avoided with elastomers used in place of PVC for refrigerator gaskets, and vacuum cleaner hoses. The number of grades employed in appliances is also being reduced, for example, in refrigerators, the number of grades used has dropped from 30 to 10 or less.

3.3.3 Dyson Cyclone Vacuum Cleaner

With most vacuum cleaners, the bag quickly clogs with dust as the suction had to pass through the bag, and a clogged bag, even with as little as 10 oz. of dirt, can cut the suction in half so the more a bag is used the less effective it becomes. Entrepreneur James Dyson set out to solve the problem. Five years and 5,000 prototypes later, the world's first cyclonic bagless vacuum arrived. When none of the major manufacturers were interested, Dyson launched his own vacuum cleaner company. Now, Dyson Cyclonic Vacuum, Europe's hottest home appliance, has come to America.

The Dyson has seven ABS plastic conical cylinders with 100,000 G's of spinning centrifugal force. No bag or filters are in the main cyclone chamber to obstruct airflow, so there is no clogging to reduce suction. Dirt and debris are literally thrown out of the airstream, which can readily be seen through the clear PP chamber. Deep inside, a lifetime Hepa Filter with Bactisafe screen kills harmful bacteria and traps microscopic particles, protecting sufferers from airborne pollutants. Ergonomically designed, the Dyson's agility and ingenious flexibility makes tedious vacuuming into a satisfying pleasure. The vacuum is perfectly balanced and adjusts automatically to different carpet heights. Although technically an upright, the Dyson outperforms the best canisters on bare floors. A 17-foot hose accessed by detaching the handle can be used to reach under furniture, or use on staircases. The collection chamber releases from the main unit for emptying.

3.3.4   High-Performance Light Fixtures

Appliance manufacturer Miele is using Ticona's Fortron 6165A4 polyphenylene sulfide (PPS) in Miele washing machine drum light fixtures for the materials resistance to humidity, repeated exposure to detergents and the heat generated by the halogen bulb. Fortron PPS, by Ticona, a Celanese AG subsidiary, is also a good substrate for overmolding with liquid silicone rubber (LSR). Installed in Miele's top of the line washing machines, the light illuminates the drum when the door is opened. The PPS gives the light socket long-term dimensional/mechanical stability in the face of the hot wash cycles and prolonged exposure to 175°C from the drum light. Fortron PPS is flameproof and meets electrical use specifications without undesirable current leakage as it has such a low level of ionic impurities. It also is a good reflector, improving lighting efficiency.

A diffuser made of polymethylpentane is placed on the light socket. To prevent the seal between the diffuser and socket from breaking when changing halogen bulbs, Miele places a silicone rubber layer on the PPS socket. This is done by overmolding the PPS socket with GE Bayer Silicone Silopren LSR 2740. The light sockets, brought to market through the efforts of Miele, Ticona, GE Bayer Silicones, and the light manufacturer Hurst & Schröder are mass produced by Hurst & Schröder, which injection molds the PPS components, and Junker & Halverscheid, which places the LSR overmold seal onto the PPS parts.

Chapter 4
Industrial End Use Applications

Keywords Packaging • RFID • Container • Stretch film • Building • Construction • Industrial • Fasteners • Pallets • Agricultural

4.1 Packaging Applications

4.1.1 Packaging Applications Introduction

Plastics remain the material of choice for packaging, increasingly substituting for other more traditional materials because they are lightweight, flexible, and easy to process. Packaging constitutes the single largest end use for plastic materials, in 2010 accounting for approximately 34% of all plastics used in the US and 37% of all plastics used in Europe. Plastics used in packaging are primarily the commodity resins: polypropylene (PP), polyethylene (PE), polystyrene (PS), polyvinyl chloride (PVC), and polyethylene terephthalate (PET). Semi-commodity nylon 6 (PA6) is an important exception, because film for food (mostly meat) packaging is a major market for this polymer. Packaging has become the major market for plastics because plastics offer better protection against spoilage/breakage, display products more attractively than do conventional materials, are lighter in weight than traditionally used paper, glass, and metal products, and offer savings in freight costs (primarily fuel economy). The need for improved product performance, faster packaging speeds and downgauging is driving the development of leading edge, cost-effective packaging solutions.

4.1.2 Stretch Hood Film Wrap

Lachenmeier, one of the leading manufacturers of end-of-line packaging machinery, has developed a new generation of stretch hood machines, which presents

D.V. Rosato, *Plastics End Use Applications*, SpringerBriefs in Materials, DOI 10.1007/978-1-4614-0245-9_4, © Springer Science+Business Media, LLC 2011

considerable advantages for end users. The innovative stretch technology, offering significant cost reduction, improved optical properties, and outstanding load stability, is positioning stretch hood as the number one choice as an end-of-line wrapping method for a wide range of industries. The blown film business for stretch hood applications is projected to increase by 22%/year through 2011 in Europe alone compared to 5.5%/year for stretch film. Stretch hoods are elastic film tubes used to wrap a stacked pallet. Stretch (mono or multilayer) film, circularly wrapped, dominates the pallet wrap market with a 67% market share. Another competitor, shrink hood film (film tubes shrunk around a stacked pallet by applying heat), is expected to decline by 2.3%/year through 2011. The total European market for pallet wrap is 1.5 million tonnes/year. While the pallet wrapping film market will advance significantly in volume terms, stretch film capacity utilization is forecast to remain below 80%, based on announced capacity increases. Thus margins will be under pressure and supply will restructure with numerous shrink hood film and stretch film players exiting the business.

When hood stretching, an undersized film hood is stretched hydraulically to a dimension matching the exact outer circumference of the load. The four grippers, which have stretched the film, travel down the load applying the film and finally securing it underneath the pallet. This bottom stretch ensures a reliable hold between pallet and load and secures the pallet for storage and transportation. Test results show that a wide range of products can benefit from this technology, not only bagged products like cement or chemicals, but also bricks, appliances, food, beverages, and other bottled goods which were initially too unstable for this wrapping method.

Lachenmeier's new stretch unit provides a controlled application of the stretch hood film with every step of the application procedure fully controllable from stretching the film, to the application over the load, and final release of the film under the pallet. A film hood with a smaller circumference than the load is wound up on four stretch units and stretched to a size matching the exact outer circumference of the load. Thanks to the Lachenmeier top stretch film unwinding system (patent pending EP 1.184.281), by means of which a controlled amount of film is unwound during stretching, thin, fragile film areas, especially on the corners of the product, are avoided.

4.1.3   Electronic Product Codes (RFID)

Wal-Mart Stores Inc. with $245 billion in revenues is the world's largest company. It is three times the size of the world's No. 2 retailer France's Carrefour. With costs a primary focus Wal-Mart has taken a major interest in RFID (radio frequency identification), where tiny computer chips on packages allow radio frequencies to monitor their location in the supply chain.

Though RFID is gaining interest among retailers such as Target Corp. and Carrefour, Wal-Mart is the leading force. The retail colossus is asking its top 300 suppliers to produce radio controlled microchip tags to be applied to shipping containers/pallets. The company began a new era in supply chain management. At that time, Wal-Mart and eight product manufacturers started testing electronic product codes (EPCs), that is RFID tags marked with an EPCglobal symbol, at select Supercenters and one regional distribution center in the Dallas/Fort Worth metroplex. EPCglobal is a joint venture of EAN International and the Uniform Code Council. It is the organization chosen by industry to develop standards for RFID technology in the global supply chain based on user needs and business requirements. Simultaneously, Wal-Mart's top 100 suppliers – plus 37 volunteers – have been working toward being live with EPCS in North Texas by Q1 2011. The company has also discussed implementation plans with their next top 200 suppliers with the expectation that these suppliers will begin EPC tagging of cases and pallets by January 2012.

The radio tags will allow every package to be identified and tracked en route to every destination through a signal sent by complex circuitry. In the supply chain application, passive RFID chips with small antennae are attached to cases and pallets. When passed near a 'reader,' the chip activates and its unique product identifier code is transmitted back to an inventory control system. Readers used by Wal-Mart have an average range of 15 feet.

As Wal-Mart is reshaping plastics packaging, adopting RFID technology, plastics processors face a daunting cost containment challenge. Thirty billion tags are required, which currently cost 30¢/tag. That price needs to be less than 5¢, and suppliers want a 1¢ tag to respond cost-effectively.

4.1.4 Transparent Paint Cans

Following their strong success in Europe, all plastic gallon paint cans are entering North America's retail paint market, with the PCC Group (The Plastic Can Company) recently introducing an injection stretch blow molded clear, one piece PET can. The company expects, based on superior cost, functionality, and manufacturability, that 15% of steel cans in the US will be converted to the PCC Group's PET design by early 2011. PCC's technology, which is available for license, molds the rim into the preform, allowing one piece PET cans, for the first time to be blown on single stage injection stretch blow machines. The snap lid is also PET. What distinguishes PET cans from other plastic versions is their resistance to both water- and oil-based paints, and their ability to be either clear or colored. Blow molded PET containers will transform paint packaging economics in the US. A round PET can is able to drop into the existing paint industry infrastructure with minor modifications.

Initial use of PET cans in North America was by Innopack in Mexico, which recently launched four sizes (1/4–4 L) for a Mexican customer. Three sizes are molded on a Husky one-stage Index SB-125 machine, while the 4 L container is molded on Aoki one-stage machines. Innopack is a leading Mexican PET beverage containers/caps molder. The company developed a see-through label and snap-on PE handle for the 4 L cans.

4.1.5 Self-Heating Container

One of the most innovative packaging solutions recently introduced is North America's first self-heating container, designed to heat liquid contents such as coffee, tea, cocoa, soups, and alcoholic beverages. Created by OnTech, Inc., the new self-heating container is a safe and easy-to-activate concept that heats up the contents of its package to approximately 145°F within minutes. The company now has 102 approved utility patent claims in the US and patents in 36 other countries, including the UK, China, and Japan covering any multichamber plastic product that must be retorted, or sterilized during a heating process. In Europe, some metal containers can self-heat and others have tried a mix of materials to create the same result but with limited commercial results. Metal has a tendency to collapse or crush during the sterilization process, when the container has to be heated to 250°F.

The OnTech self-heating container has three main parts, the six-layer plastic container, an inner plastic cone, and a water holding activation 'puck.' The inner cone holds the crushed calcium oxide or 'quicklime,' used in the heating process, and the outer container body holds the beverage product. The puck, which holds the water to react with the calcium oxide when activated, fits inside the cone, and a tamper-evident foil membrane seals the operating mechanism. To activate, the consumer removes the tamper-proof foil and firmly presses down on a button on the container end. This releases the water contained in the puck into the calcium oxide within the cone and begins a thermal reaction that heats the inner cone to 250°F. This heats the beverage in the container to 145°F in 6–8 min, while the lukewarm outer shell remains a comfortable temperature to hold. The beverage will stay hot for at least 20 min and warm for up to an hour. The high barrier disposable container is shelf stable and does not require refrigeration prior to opening.

The six-layer plastic container consists of four FDA approved plastic materials: polypropylene, an EVOH oxygen barrier material, a tie layer, and a regrind layer. The result is a strong, safe container that may be used with many different types of products. The container is designed to withstand normal manufacturing procedures for the can filling industry and can therefore survive the retort process as required by the FDA for canning certain food products. The OnTech container is a standard 16 ounce size and will hold 10 ounces of product (the remaining 6 ounces is occupied by the inner heating cone). It is designed to accommodate standard fill lines and is suitable for hot fill, retort or ionized pasteurization. A smaller, standard 12 ounce OnTech container is planned to hold 7 ounces of product, suitable for baby formula, sake, specialty drinks, etc. The container can also be molded into custom shapes to fit specific needs (soup bowl, mug, baby bottle). Each of the containers will bear the label 'Powered by OnTech,' to remind consumers of the technology behind the product. The company is also working on the next iteration, a self-cooling plastic container that would chill the liquid, such as beer or soft drinks.

4.1.6 Elastic Shrinking Stretch Film

Easiwrap, the signature product from Easiwrap International, is made with a new technology that transforms conventional thinking about stretch film. The engineered palletizing film is a stretch film that acts more like a shrink film. Elastic like Easiwrap's stretch 'memory' is pre-programmed at the manufacturing stage through a unique technique that makes it seem more like a shrink film. The proprietary film and manufacturing process was the brainchild of Irish engineer Phillip Dorn, who invented the technology to make the blown film and the mix of materials that go into it. When the corners of the Easiwrap film are tugged, it shrinks to fit as tight as an elastic stocking. A slight tug in the corner orients the molecular structure, starting it shrinking for 20–30 min after it is wrapped around a skid. Since reactivity of the film starts with that tug, the wrap can be applied loosely to a pallet load then tightened making the pallet wrapping operation easier and quicker. With its 300% bilateral stretch, the film can conform to irregular dimensions on the pallet and reduces film usage, reducing material waste, increasing productivity, and making the cost per pallet more competitive. By forming a tight fit the film resists nicks and punctures providing excellent product protection from damage in transit. Customers include window fabricators, building supply firms and food/beverage distributors.

4.1.7 Rectangular Beverage Bottles

Gatorade's 'Thirst Quencher' brand in now packaged in a redesigned container, the first hot filled rectangular PET gallon bottle to reach the market domestically or internationally.. The new bottle, created by Owens-Illinois (O-I) Plastics Group, replaces the round 'space-consuming' shape previously used. The redesigned gallon bottle is easier to handle, allows retailers to display more product on store shelves and customers can store the bottle more easily in the refrigerator door or other tight places.

Considering that a gallon weighs 8 pounds, the container can be cumbersome and difficult to grip. To improve handling, O-I designed a bale handle for the bottle to give users, particularly children, something to hold on to when taking the product from the refrigerator and pouring it into a cup. The custom HDPE Uni-Pak bail handle, which is applied to the bottle prior to filling is supplied by PakTech. In addition to creating a rectangular package that was easy to handle and pour, Gatorade required a bottle that could be filled on existing bottling lines. This necessitated positioning the handle so that the neck could still be gripped during the filling operation. The bottle is molded with recessed hand grips on either side that and ribbed panels that, together with the recessed grips, make the heavy bottle easier to maneuver while preventing paneling during hot filling at temperatures up to 180°F. From initial concept of the container to production, the new monolayer PET bottle took 6 months to complete.

4.2   Building and Construction Applications

4.2.1   Building and Construction Applications Introduction

Building and construction (B&C) is the second largest end use market for plastics. The building and construction industry uses plastics for a range of applications from insulation to piping, window frames to interior design. It is plastics durability, strength, resistance to corrosion, low maintenance and aesthetically pleasing finish that ensures their continued popularity in the sector. Most of the volume usage is processed by extrusion, such as for piping, conduit, wire insulation, siding, reservoir liners, erosion control netting, or architectural sheeting. As in packaging, plastics have displaced such traditional materials as wood, glass, and metal, based on improved performance and lower cost. Considering the volumes involved, the marketplace tends to favor commodities wherever possible.

4.2.2   Fiber-Based Composite Building Products

Kadant Composites Inc. is using a paper sludge by-product mixed with plastic to produce extruded deck boards. Biodac, a Kadant patented product, is a synthesized recovery of short and long cellulose fibers, precipitated calcium carbonate and kaolin clay that would otherwise go to a landfill. A Kadant sister company produces a granule from paper sludge, a paper recycling waste product, which has found application in a variety of industrial uses particularly as a carrying agent for fertilizer. In the process of extruding that granule, a very microscopic granule is created that was too small to be used in any of the applications that the sister company was selling into, and it was being hauled off to landfill. Now instead, the material is mixed with recycled polyethylene and rice hulls to make GeoDeck.

The Biodac granules, which are uniform in size and shape, and mix extremely well with HDPE, are a critical element in the production of a composite building material being used to produce Kadant's extruded Geodeck deckboards. Unlike wood fibers, which contain lignin and tannin that inhibit resin bonding, Biodac is low in those chemicals and compounds more readily with plastics. Kadant developed GeoDeck deck profiles, hollow tongue-in-groove boards of HDPE with up to 30% Biodac plus additional fiber from rice hulls. The composites are approximately one-third plastic, one-third mineral, one-third fiber. Up to 70% Biodac could be used, but these high levels are too abrasive for the compounding extruders. Whiteners and brighteners from the paper boost the effect of colorants in GeoDeck, making it less subject to fading than wood–plastic composites. The product has greater dimensional stability and less moisture absorbance than competing composites. That granule is the key ingredient that gives Geodeck its fade resistance as well as forming a very, strong product.

GeoDeck sales have grown from under $2 million in 2007 to more than $15 million last year. Kadant Composites has developed other new building products containing Biodac, including corrugated roof tiles that look like red clay, and flat, square tiles that resemble slate and is also introducing new Geodeck products.

4.2.3 Flexible Mouldings

According to Style Solutions Inc., customized construction products and large builders will drive the growth of urethane millwork in 2010. The firm has created more than 300 customized products in the past year for builders across the US. The building and construction industry has seen unprecedented advances in the acceptance and use of urethane millwork on both exteriors and interiors of commercial and residential buildings. Just 5 years ago, wood moulding profiles were predominate with few engineered plastic ones seen. Today the reverse is true with low maintenance products, such as urethane, composite, and plastic mouldings prominent in all regions of the country. Urethane millwork, which stands up to high temperatures in the summer and frigid snowy weather in the winter months, is being increasingly used with other products such as doors, windows, and vinyl siding.

Style Solutions recently introduced a new line of decorative moulding pieces that can be manipulated to fit most any curve or bend, making it easy for homeowners to add moulding accents to their homes. The new 'Flexible Mouldings' with the firm's most popular moulding profiles allows consumers to trim radius walls, curved stairways, and arched entryways. Each piece, made of tough, high-performance urethane, vinyl, or styrenic elastomer materials, can be bent more than one way, making it possible to 'snake' flexible moulding down a staircase or curve it completely around a column. For interior and exterior applications, the mouldings resist insect infestation, rotting, decay, and splintering. The closed cell structure used also prevents water penetration and absorption. Each moulding piece comes pre-primed. The unique flexible pre-primer topcoat stretches with the moulding to minimize cracking. Once installed, the flexible mouldings can be painted, stained, or faux finished.

4.2.4 Artistically Designed Floor Coverings

Bayer's Artwalk marries the practical with the beautiful. A successful marketing concept, Artwalk innovative coatings is the brand name for a new line of polyurethane raw materials from Bayer MaterialScience for creating individual and highly decorative floors for indoor and outdoor applications. Among the floor coverings formulated with Bayer's Desmodur (blocked polyisocyanates) and Desmophen (polyether, polyester and acrylic polyol) are cast resin coatings or terrazzo-look granular rubber flooring. One design advantage shared by these systems is that they can be laid on seamlessly on site and very intricate designs can be applied over the surface and particularly to the curved surfaces at columns, sharp corners, or irregular lengths of wall. It is also possible to create parquet-type inlays and logos. The materials offer a variety of unique design options to produce one-of-a-kind floor surfaces that can impart a clear identity.

With the large assortment of colors available for both systems, the floor design possibilities are almost endless. In addition to the cast resin coatings with their myriad design possibilities, the Artwalk line also includes colored granular rubber floor coverings. Depending on the system selected, a combination with various special effect additives such as aluminum particles, glitter, wood pieces, or chips is also possible. Both coating systems can also be formulated to meet specific stringent functional requirements, i.e. high abrasion resistance, mechanical strength, color-fastness, resistance to chemicals, crack bridging, and ease of maintenance. They adhere to almost any substrate. Both coating system surfaces deaden the sound of footsteps. Another important characteristic is their high yellowing resistance, which ensures a long-lasting, pleasing appearance of the floors.

Hard-wearing granular rubber floors combine optimum creative freedom with maximum wear resistance. A mixture of rubber granules and a polyurethane binder (Desmodur/Desmophen) makes it possible to produce seamless graphic designs. The material can be used to achieve smooth and gradual color transitions, as well as interplays of light and dark. Depending on the composition of the rubber granule/binder mixture, the floor covering ranges from soft through flexible to ductile. The polyurethane granular rubber flooring, by its very nature, can be used to level out significant irregularities in the substrate. No screed is required under granular rubber flooring.

Bayer is targeting both the public and private sectors with Artwalk, in particular showcase buildings such as chain stores, restaurants, sports stadiums, hotels, banks, insurance offices, fitness studios, and film studios. A fixture of the successful marketing concept is the Bayer MaterialScience offer extended to architects and planners to coordinate all aspects of project implementation. Project teams comprising designers, formulators, and applicators can be assembled for each property and custom solutions developed upon request.

4.2.5 Sandwich Construction Plate

British Canadian company Intelligent Engineering Ltd., in cooperation with Elastogran GmbH, a member of the BASF group, brings the marine and civil engineering industries a unique concept in steel construction. The patented Sandwich Plate System (SPS), in which polyurethane elastomer is injected between two steel plates represents a major advance in steel engineering, in all applications where giant steel parts are used. Wide applications exist, from shipbuilding to repairing steel-plate structures such as bridge decks, railcars, and ship decks, hulls, and hatch covers. SPS also offers benefits for the construction of sport stadiums and arenas.

Licenses, for the use of SPS technology, have recently been awarded to major companies in Europe and North America.

The huge steel elements used in shipbuilding or bridge structures are exposed to enormous mechanical stresses. In conventional steel construction steel plates are reinforced with welded-on longitudinal ribs to maximize stiffness and minimize deformation in a process that requires appreciable labor. The welds of these ribs are also typically the first point of attack for corrosion and fatigue. SPS is a much more effective, elegant method of reinforcing large steel components. The elastomer is simply injected between the two steel plates in place of the welded longitudinal rib reinforcements. The result is a stiff steel-polyurethane-steel composite, which dramatically reduces construction complexity and removes the majority of fatigue and corrosion prone details when compared to stiffened steel. SPS structures are lighter and faster to build than conventional maritime and civil engineering structures and also offer built-in protection against fire and vibration. Other benefits include acoustic insulation and improved dimensional accuracy.

4.2.6 Secure Window Film

CPFilms Inc. is the world's largest manufacturer of solar control and safety window film for automotive and building applications. The company, with DuPont Teijin Films (www.dupontteijinfilms.com), is jointly developing and marketing LLumar Magnum safety and security protective films made with Mylar from DuPont Teijin Films. Damage and injury from projected glass can be extensive. The alliance brings technology and expertise to commercial and residential markets to help control or prevent damage and injuries through expanded application of well-designed safety and security window film.

LLumar Magnum safety and security films provide a protective sheath on the interior of glass that helps hold fractured shards of glass together should a break occur, whether from natural disaster, terrorist act, vandalism, theft, or accident. The micro thin film is composed of multiple layers of strong, clear, special grade Mylar polyester film and metallized coatings bonded together with a unique adhesive to form a shield that holds glass in place. The film also resists scratching and provides solar protection by screening out radiant heat, and damaging ultraviolet rays, while deflecting harsh glare. The security films are available in five thicknesses: 4, 7, 8, 11, and 15-mil. The 4-mil films are considered to be safety films rather than true security films. The 7-mil film, the standard security product, offers resistance to breakage. The 8-mil films offer both solar control as well as security protection, while 11-mil films are recommended for high-risk areas, such as store windows, and offers some level of protection in case of explosions, where the greatest danger to building occupants is flying glass. The 15-mil films offer the ultimate in protection in highest-risk areas, and give significant protection from explosions.

4.3 Industrial Applications

4.3.1 Industrial Applications Introduction

Industrial is one of those categories that cover a lot of territory – everything from heavy equipment to nuts and bolts. Caterpillar, the market leader in heavy equipment, says its machine sales in North America for 2010 rose 19% from a year earlier, the biggest gain since 2006. Foreign sales were down 6%, but worldwide infrastructure construction is pushing up sales of equipment globally by 7%. Others serving the construction equipment segment are also experiencing positive growth. Ingersoll Rand Co. reported strong sales of generators, compressors, and Bobcat construction equipment. In the nuts and bolts end of the industrial market segment, industrial fastening systems are projected to grow 5.6% through 2011, fueled by a healthy increase in durable goods production and construction. Strong sales of higher-performance specialty products such as aerospace-grade fasteners will experience the fastest growth as manufacturers replace commodity items with custom designs. Among the top 45 players in this market are Illinois Tool Works, Textron, Emhart Fastening Technologies, SPS Technologies, Alcoa, and Park-Ohio.

In between heavy equipment and fasteners, industrial conglomerates include Honeywell International Inc., a $22 billion manufacturer of products for both industrial and residential markets. Honeywell makes controls for heating, cooling, ventilation, humidification, industrial process automation, and video surveillance, as well as access control equipment, security/fire alarm and industrial safety systems, and home automation systems in its automation and control products segment,. Honeywell's 2010 earnings grew between 9% and 17% in all four of its operating segments. ITT Industries is a diversified industrial manufacturer whose Fluid Technology and Motion & Flow Control units produce switches, valves, pumps, and fluid handling systems. ITT expects improving conditions in its key markets to provide solid growth opportunities. Tyco International's Engineered Products & Services business, with 2009 revenues of $4.7 billion, saw its market segments grow in 2010. Products for industrial process equipment and water/wastewater systems include such things as valves, actuators, controls, and thermal control systems.

4.3.2 Energy-Saving Freezer Apparatus

Creative Plastics & Design (CPD) has designed and manufactured a plastic deck pan (patent pending) that replaces the industry standard stainless steel pan used inside refrigerated/frozen food and beverage display cases. The deck pan created for refrigeration systems company Hill Phoenix serves as a barrier that separates and protects perishable foods from the freezer's cooling/self defrost cycle. The new pan, made from HDPE, features a blow-molded design that offers greater durability and food safety than the original stainless steel pan, while cutting processing costs by 67% and weight by 30%.

The CPD design features supporting ridges on the pan's top and bottom surfaces. Adding rigidity and durability to all sides of the pan, the new design enables it to meet requirements for static, side, axial, and torsional loads as well as for impact strength. According to CPD, the HDPE resin selected was the only plastic material that could meet impact strength requirements without becoming brittle and weak during testing at refrigerator/freezer operating temperatures. The new deck pan design also delivers an insulation factor of R-3 without using any foam backing or covering. The new display case deck pan has earned National Sanitation Foundation (NSF) approval permitting contact with food and can serve not only as a barrier but also as an attractive display rack, fitting existing refrigerated cases.

4.3.3 Plastic Pallets

With stricter hygiene standards, concerns about pests, and greater design features, the plastic pallet market is enjoying global demand growth of more than 10%/year At least four plastics processes are seeing extensive use: injection molding, rotational molding, single- or twin-sheet thermoforming, and structural foam molding. Responding to demand growth, Kiga GmbH has developed a light weight plastic pallet that is cost competitive with untreated wooden ones. These single use pallets made of recycled polyethylene or polypropylene are frequently used to ship products to the European Union, China, Japan, Australia, and other countries that do not allow the entry of untreated wooden pallets for fear that insects may be harbored in the wood. Single use pallets of recycled plastic typically cost €5–€6, while those using virgin material designed to meet extreme hygienic standards and high-strength requirements can cost as much as €50.

Plastic pallets are more durable than wooden pallets in multi-trip applications where plastic pallets can often survive 100 trips, compared to 10 trips for wood. Kiga uses recycled PET for some pallets as the material is stronger than PE and also inserts iron bars in others to improve load strength. The company, which both injection molds and compression molds pallets, recently started thermoforming a collapsible version for Chrysler. Collapsible versions require less storage space after they have been emptied, an important point for many automotive OEMs and other manufacturers. Thermoforming allows the use of much less expensive tooling – in the range of €10,000 vs. €250,000 for injection molds and the tooling can be made in 3–4 weeks. But injection molded pallets can be more precisely manufactured, an important contrast to wooden pallets, which may have knot holes and eventually sag, warp, or bend. Injection molded units show little change over time, and this is critical in many firms where increasingly it is robots that are unloading pallets.

4.3.4 Sterile Disposable Flasks

Quality labware manufacturer, Nalge Nunc International (NNI), an ISO 13485 certified world leader in the production and supply of laboratory plasticware for use in research and educational laboratories, has reduced product and manufacturing costs in their latest line of disposable flasks. The new Nalgene sterile disposable flasks, ideal for shaker and suspension cell culture, media preparation or storage, combine an innovative lightweight design with the versatility of Eastman Chemical's PETG plastic, Eastar copolyester 6763, to deliver a cost effective, completely disposable product. Lower material and processing costs allow NNI to offer the new Nalgene sterile disposable flasks at roughly 60% of the cost of competing brands. The flasks are ready-to-use, in gamma sterilized individual packages. They are also free from the yellowing typically associated with other polymers when exposed to gamma sterilization.

Designed for disposability, amorphous copolyester enabled NNI to mold an extremely light weight, thin walled product with the performance integrity of its heavier polycarbonate counterpart. The readily disposable flasks weigh only a fraction of that of other materials including glass or polycarbonate, making disposal more cost effective. Disposability greatly reduces the risk of cross-contamination, and the flasks can be cleanly incinerated, allowing laboratory professionals to realize process efficiencies from initial use to final disposal.

4.3.5 Mechanically Efficient Microplates

Tecan, a leader in the Life Sciences supply industry, is working with Corning Inc. to offer an innovative microplate handling system for low volume microplate assay applications. While small liquid volumes can be dispensed, the ability to work with low volumes (low ml or nl) for various assay and molecular biology applications has been severely limited by rapid evaporation, especially at the elevated temperatures typically required for many types of assays. For these applications, Tecan and Corning Life Sciences have assembled an innovative, new microplate handling system, which combines Tecan's Te-Lidder COC plastic microplate lid handling module and Corning's Robolid technology. The system provides an ideal solution for sealing, unsealing, and resealing a variety of commonly used 96 and 384 well microplates.

Tecan's Te-Lidder is a robotic workstation accessory designed for use with Corning's Robolid, a novel microplate seal that incorporates a silicone capmat into a standard polystyrene plate lid. This combination provides each well on a standard plate with a liquid tight seal in a format that is automation compatible. Tecan's Te-Lidder is specifically designed for use with Robolid technology on Tecan Workstations and TRAC Robotic Systems. This patented technology is designed for easy and quick removal of tight sealing Robolids from a wide variety of 96 and 384 well plates.

4.3.6 Aerodynamic Truck Brackets

To reduce cost and weight, RTP Company teamed with Volvo to develop two types of Volvo truck brackets, a pair of L-shaped faring brackets used to mechanically fasten the truck cab's aluminum grab handle to the truck cab side, and brackets designed to hold a roof mounted satellite dish. Requirements for the new injection molded grab handle brackets, which are prominently located on the truck, included a quality surface that could be top coat painted without primer to control cost. The grab handle also had to pass testing in which it was subjected to a 250 lb load, necessitating a minimum bracket tensile strength of 20,000 psi. In addition, the material replacing the formerly used metal had to display shrinkage rates corresponding to those of the existing tool. The satellite bracket required a flexural modulus of at least 1.2 million psi, and since it did not require painting, the material would have to withstand chemical and ultra violet radiation exposure of typical highway conditions.

The truck brackets, formerly of banded steel, were injection molded from RTP's 4700 Series polytrimethylene terephthalate (PTT) compound, which contains both a UV protection package and glass fiber reinforcement. PTT is a semicrystalline polyester, closely related to the more common thermoplastic polyesters PET and PBT, but is typically a less expensive material. RTP Co.'s new RTP 4700 Series balances the physical properties of PET with the processing characteristics of PBT. The PTT compounds offer strength, stiffness, and high heat deflection temperatures of PET; low melt/mold temperatures and wide processing window of PBT; basic polyester benefits of dimensional stability, electrical insulation, and chemical resistance; and cost savings.

4.3.7 High Tech Fasteners

Swiss-based fastener maker, Icotec AG (Innovative composite technology), produces fasteners for the aerospace, automotive, and medical industries using a novel compression flow molding (CFM) process that delivers net shape thermoplastic composites with strength per unit weight competitive to, or exceeding machined steel, aluminum, and titanium. The innovative technology developed by Icotec is used by the company to produce high strength light weight fasteners. With its CFM process the company is able to put up to 62% carbon fiber by volume into thermoplastics such as PEEK, creating a high-strength, abrasion-resistant composite.

The lightweight nuts and bolts thus produced sport ultrahigh specific strengths of up to 2.4 million psi/lb/in^3 This contrasts to aluminum and titanium with specific strengths of about 0.2 million and 1.2 million psi/lb/in^3, respectively. The CFM process feeds pultruded composite rods into a closed preheating cavity. From there, the rods move to the mold and are formed to shape under pressure with helically reinforced threads. The high-volume CFM process produces fasteners in a single step without the need for any posttreatment. Even with high fiber loadings, fastener threads can sustain high mechanical loads without shearing, thanks to a transitional fiber orientation zone that transfers forces away from the threads to the fastener core. In contrast, conventional thermoplastic processing techniques do not permit extremely high fiber content nor can they process long fibers without damaging them.

4.4   Agricultural Applications

4.4.1   Agricultural Applications Introduction

Agricultural plastics continue to account for just 1.9–2.6% of the total plastics consumed, in the United States and Europe, but they have a vital role to play. Plastics based agricultural systems provide effective solutions to crop growing: in arid regions, for example, plastics piping and drainage systems can cut irrigation costs by one to two-thirds while as much as doubling crop yield. The agricultural sector is also a significant application for plastic film used for soil conditioning, crop cover and grows tunnels. Elsewhere plastics are used for pesticide and fertilizer packaging, and to produce farm implements and equipment. Plastics also plays a growing role in animal husbandry displacing traditional materials as wood, and metals for products such as feed troughs, watering tanks, grain bins, and even horseshoes.

4.4.2   Farmland Demining Equipment

Land mines pose a lingering threat beyond risks to life and limb. Strewn across farmlands, mines can cause famine by rendering fields unusable. In cooperation with Caterpillar Inc., Countermine Technologies AB has developed Oracle (Obstacle Removal and Clean Land Equipment) a complete unique demining and land reclamation system.

Oracle based on Caterpillar's off the rack loaders costs 1/100 as much as a military system to set up and run. The system consists of Spitfire, a mine clearing tool, which clears mines placed both in and on the ground to a depth of 45 cm, will in excess of the UN 20 cm requirement. The patented tool destroys a mine before it detonates. Mounted on a caterpillar, the Spitfire tool can clear up to 15,600 m²/h with a clearance rate of 99.6%. Girded by steel plated armor, yet employing lightweight RIM, SRIM and related reinforce plastic composite parts for interior weight reduction and fueled by a tow along power plant, it uses its massive sharp toothed rotor to shred mines before they go off, or if not, to simply smother the explosion. Oracle also tills the ground as it traverses the field so that the land can be used directly after the mine clearance operation. Farmers welcome Oracle with open arms, and typically throw the Oracle team a party.

4.4.3   Engineered Horseshoes

Senior & Dickson, a UK-based molder and toolmaker, has produced the first ever plastic horseshoe. The shoe, produced on behalf of Lightfoot Horseshoes Ltd. is a two-shot injection molding using Nylon 6.6, and a TPE compound. In addition to the two plastic compounds, the horseshoe contains a small steel strip, which is inserted in the mold.

The new plastic horseshoe, the Jameg Sprinter, offers many advantages over traditional steel or aluminum. Designed to be glued to the hoof, the shoe includes a series of removable clips to aid in fitting and a unique rubber lift on the foot side surface to help create the correct acrylic adhesive thickness. No nails are required, an important factor as nails, which damage the hoof can limit the number of times an animal can be reshod, and in the case of racehorses, the frequency that it can compete. Equally important, because a horse's hoof is effectively a polymer, the nylon shoe base produces both strength and flexibility, working in harmony with the hoof, not against it. Once attached, the shoe becomes part of the hoof, allowing it to expand and contract naturally. Ultralight weight and slip resistant the Sprinter can be fitted with studs and used on horses involved in a range of disciplines. They are less likely than steel shoes to cause injury to a fallen rider, and are ideal for brood mares, horses at rest, and those with poor or badly broken feet.

4.4.4 Ag Equipment Large Parts

The John Deere 7000 Engine enclosure is the first use of co-injection to produce cost effective large parts from engineering grade material in thick walls – 6 mm to 8 mm. Molder, Bemis Manufacturing Company a leading global manufacturer of diverse proprietary and custom plastic products, uses coinjection molding, allowing regrind to be used in the core to reduce costs. The hood and side panels of the John Deere 7000 engine enclosure consist of a skin of polycarbonate/polybutylene terephthalate, and a core of regrind ABS. This construction yields parts with excellent high impact strength and very good part stiffness. The hood also includes a one piece front grille frame allowing the entire grille and light system to be mounted to the frame as a module. Each hood assembly contains 15–20 pounds reclaimed material. A polyurethane topcoat is applied to the cosmetic body parts of the tractor engine enclosure.

Three mold makers were involved to produce the required molds for the tractor engine enclosure Triangle Tool Corporation and Tooling Technologies, Inc., both of Milwaukee, WI, and CDM Tool & Manufacturing of Hartford, WI. The tooling, made by Triangle Tool, includes two large core pulls, four sets of expanding lifter units, and seven collapsing core sections.

4.4.5 High Durability Irrigation Filters

Systems Europe Srl, a Rome-based European subsidiary of Toro Irrigation, has developed a large agricultural filter that has high burst strength and excellent environmental resistance. Taking over a year to develop, the new cartridge filter, which is positioned between the inlet and outlet systems of the irrigation's main pipeline serves to clean the water, eliminating dust and other particles and ensuring a smooth, constant, and uninterrupted flow of water. The filter consists of three parts, the main body, the thread, and the cylinder.

The company chose Zytel for the irrigation filter as the nylon resin has excellent mechanical and burst strength, and outstanding resistance to ultra-violet rays, water and a range of chemicals and fertilizers, even at temperatures of 80°C in hot summer sunshine. The filter in product literature is described as resisting 10 bar pressure, and Toro/Irritrol says it has resisted extensive trials at up to 24 bar pressure, providing a safety margin of more than 2:1. Toro/Irritrol is targeting this high quality filter at mid-level farmers worldwide.

4.4.6 Ag Pesticide Recyclable Containers

More than 7.3 million lb of crop protection/pesticide containers were collected for recycle in 2010. Thirty member companies, through the Ag Container Recycling Council (ACRC), fund pesticide container recycling programs to collect the HDPE plastic containers throughout the US. If not managed properly pesticide containers can pose an environmental hazard. Farmers were previously burying the containers on the farm, sending them to the landfill, or using them improperly to store other chemicals. Faced with growing concerns about the handling of the empty containers, farmers and pesticide producers saw value in proactively creating a system to address the environmental concerns surrounding pesticide jugs and bottles before the government took legislative action.

One of the most successful and innovative recycling programs in the country, the ACRC recovers more than 30% of the 30+ million containers produced each year. The ACRC pays for the collection and recycling of used containers of both council members and nonmembers. Containers must be triple-rinsed or pressure washed before being collected. The empty bottles are inspected when they are returned to ensure that they are properly cleaned. Some collection sites are operated through public agencies, and many are located at crop protection products dealers. Contractors often grind the containers at the collection site for easier transportation. Recycle applications are limited to nonconsumer products that do not pose a significant risk of human exposure including the inner core of marine pilings, and drainage pipe.

Chapter 5
Transportation End Use Applications

Keywords Transportation • Automotive • Thermoplastic • Thermoset • Adhesive • Glazing • Carbon fiber • Aerospace • Marine • Boat

5.1 Automotive Applications

5.1.1 Plastics and Transportation

The transportation industry is the largest consumer of engineering plastics, both as thermoplastics and as thermosets. This industry is also the testing ground for new products and processes, and a leading indicator of plastics usage in other industries. The safe and cost-effective transportation of people and goods is vital to the global economy. As modern modes of transportation have evolved to meet increasing demands for safety, environmental protection, and speed, the use of plastics in transportation manufacturing has grown considerably. All forms of transportation require energy to operate and fuel represents a significant part of operating costs. Reducing vehicle weight be it cars, airplanes, boats, or trains can cut fuel consumption dramatically. Used in the service of vehicle exterior parts, plastics are durable, do not corrode, and require little maintenance. They allow a unique freedom of design and the parts are fast and economical to manufacture. In vehicle interior applications, the various plastics fittings, such as, flooring, seats, dashboards, and paneling maintain their attractive appearance and are easy to clean. Durable, lightweight, and corrosion-resistant plastics offer fuel savings, design flexibility, and high performance at lower costs to designers facing today's complex transportation needs.

D.V. Rosato, *Plastics End Use Applications*, SpringerBriefs in Materials, DOI 10.1007/978-1-4614-0245-9_5, © Springer Science+Business Media, LLC 2011

5.1.2   Automotive Applications Introduction

Plastics play a major role in today's vehicles. The automotive industry is the single largest end user for many engineering plastics, such as nylons (PAs), polycarbonate (PC), acetal (POM), or modified polyphenylene ether (PPE). It is also an important market for commodity polymers (e.g., PP, PE, and PVC). Auto makers choose plastic parts for their durability, corrosion resistance, toughness, ease of coloring and finishing, resiliency and light weight. Automobile designers also discovered that plastics solve one of their most complicated design problems: what to do with the fuel tank. Using plastic gives them the freedom to fit tanks into the overall concept rather than designing around the unwieldy but essential part. More plastics, by volume, than steel are now used in today's cars for a myriad of components. At the end of a vehicle's working life, plastics components can be recycled or the energy can be recovered through incineration. Plastics versatility aids the automotive industry in meeting ever more stringent requirements in terms of economical performance, safety, comfort, and environmental considerations. Recent industry estimates expect the use of engineering thermoplastics in exterior automotive applications to grow by 4.8% annually through 2011, which would equate to a total of more than one billion lb/year. By 2012, the figure is expected to climb to 1.3 billion lb.

5.1.3   Carburetor and Air Intake System Firsts

Solvay Advanced Polymers, LLC, a subsidiary of Solvay America, Inc., produces high-performance polymers that are used in a wide range of demanding applications including automotive. Several recent automotive under-the-hood application developments have expanded the use of Solvay's high-performance polyphthalamide (PPA) in this market sector.

Carburetor expert, Willy Krup, developed a unique fuel pickup tube for use in oval track racers, where the nearly constant left turns otherwise push fuel to the right side of a carburetor's float bowl, reducing flow to the jets that feed the cylinders on the left side of the engine. Krup innovated a fuel pickup tube that returns this errant fuel to the fuel metering block, assuring constant flow – and consistent power – through even the fastest turns. When the large amounts of machining required to make the parts in aluminum was demonstrated to be cost ineffective, the designer turned to plastic for an affordable composite solution. Amodel AS-1133 HS, a heat stabilized, glass reinforced PPA that combines a high heat deflection temperature with high flexural modulus and high tensile strength proved to be up to the task, when molded nylon failed to stand up to the heat/cool cycles and corrosive alcohol fuel encountered in the race environment.

In other areas, a new coating designated IC-105, developed by TechLine Coatings, Inc. of Murrieta, CA, is expanding the window of opportunity of PPA in high-temperature under-the-hood applications. The coating provides thermal reflectivity when applied to the surface of parts made of PPA. The company specializes in coatings for the high-performance automotive market. The new product based on technology used in its IC family of coatings for metals is the first designed by Techline specifically for application to plastic. In testing, PPA sample plaques coated with a thin film of IC-105 were subjected to direct contact with a heat source at 600°F. After extended exposure, the temperature of the plastic in the samples remained at approximately 400°F.

The 190–200°F differential is significant for plastics such as polyphthalamide, polyamides, or amide-imides that are intended for use in components of an automobile power train. While these plastics have excellent qualities, almost all begin to lose their primary characteristics above 500–550°F. Advances in engine technology, fuel management, and emissions requirements continue to drive up under-the-hood temperatures. Where these types of plastics were adequate a few years ago, today that is no longer the case. A typical example is the engine's air-induction system where materials such as Amodel PPA are replacing aluminum. In contact with high-temperature air, the end tanks surface of the turbocharger air cooler housing can undergo oxidation that can contaminate the turbocharger system. The IC-105 coating technology improves the window of opportunity for Amodel grades to be used in higher-temperature applications, not only as a replacement for aluminum, but as an alternative to much more expensive polymers such as PEEK (polyetheretherketone).

Solvay's PPA has also been specified in the first known composite automotive performance aftermarket manifold. Amodel A-6135 PPA was specified for the new LSX air intake manifold engineered and designed for performance parts provider Fuel-Air-Spark Technologies (FAST). FAST wanted a part that outperformed the stock nylon manifold, but was not subject to the weight and heat penalties imposed by an aluminum replacement. The LSX manifold will replace the stock air intake that comes with the eight-cylinder engines in the Chevrolet Corvette, and Camaro as well as Pontiac's Firebird. Using A-6135, designers were able to increase airflow by 25% and make the runners longer to maintain low end torque, thereby adding 20 hp to the 360 available from the V-8s the LSX is designed to fit. Turbo or superchargers elevate manifold pressure so material strength is also important. Flexural modulus of the A-6135 is at least 20% greater than that of nylon 6 or 6/6, and offers greater resistance than nylon to tensile creep, which could cause air leakage at the manifold's gaskets. A-6135 also has the chemical resistance required in under-the-hood applications and is more resistant than nylon to degradation from new long-life coolants.

5.1.4   LESA Part Adhesive Technology

Dow Automotive's new LESA adhesive (Low Energy Surface Adhesive) technology has enabled the division to introduce innovative and cost-effective construction of front-end systems. The new approach addresses industry needs for improved quality and performance while cutting costs and reducing weight. LESA is a two-part system that cures at room temperature in the presence of oxygen from the air. Open time after mixing can be varied from 2 to 15 min. Bonded assemblies can be handled after a few minutes, though full cure takes 24 h.

The new front-end system produces an adhesively bonded two-piece construction front-end carrier that allows materials and manufacturing processes to be optimized with reduced manufacturing investment. LESA can bond similar or dissimilar substrates like plastic to plastic or plastic to metal without the need of a primer. The modified front-end carrier lowers the mass of metal and plastic to reduce cost while LESA adds stiffness and overall durability by reducing fatigue and failure of spot welds and other traditional fasteners. Structural noise is also reduced using the adhesive by limiting vibration caused by mechanical fasteners and spot-weld fatigue improvement. This new approach integrates advanced adhesive technology with unique design of a plastic (polypropylene) molding and metal reinforcement. The adhesive technology enables a continuous bond between the plastic and metal part for added stiffness without pre-treatment of the substrates. The design incorporates a closed box section not possible with the traditional overmolding technique, enabling maximum stiffness with minimum weight.

5.1.5 Optical Glazing

One of the leading developers of optical plastics for automotive glazing is Freeglass GmbH, a joint venture between glassmaker Saint Gobain and plastics processor Schefenacker, a specialist in exterior mirrors and lighting. In addition to the rear sidelites for DaimlerChrysler's Smart Fortwo coupe and Smart roadster, the business is also producing four polycarbonate glazing components, including a 1 m² sunroof for the 4-door Smart Forfour sedan.

Freeglass has invested in Krauss-Maffei swivel platen machines for multi-component injection molding. The Krauss-Maffei expansion compression molding process, combined with swivel platen technology, makes it possible to produce large area parts virtually free of internal stresses. Post mold, the surface of the PC parts are coated for ultraviolet and scratch protection. The plastic sunroof module is 7 lbs lighter than an equivalent glass module when the frame is integrated. This weight reduction lowers the vehicle's center of gravity, helping to improve driving characteristics. The technology also opens the opportunity for innovative design and the integration of various functional elements such as lighting systems.

5.1.6 Class A Roof Appliques

Roof appliqués are being molded from DuPont Rynite PET thermoplastic polyester resin for the Saturn VUE sports utility vehicle. The appliqués long curved exterior parts are attached to the metal space frame above doors and behind rear side windows. While providing a finished look, they also help channel rainwater away and provide a rigid, smooth sealing surface for door edge weather stripping.

Saturn specifications for the appliqué call for a Class A finish A, a demanding surface quality specification for parts that are to be subsequently painted. The painted Rynite gives the gloss required to match the other exterior surfaces of the vehicle at significantly lower cost than the alternatives. SMC or metal parts would require extensive finishing steps prior to part painting, while Saturn's Rynite PET appliqués require only minimal surface preparation. The material with its dimensional stability, low coefficient of linear thermal expansion and warp resistance also meets the consistent fit requirement. The appliqués, which are subject to high thermal load in sunlight, resist heat-related expansion and warping. Atmospheric humidity also has no effect on the dimensions, strength, and toughness of the material. Rynite RE5309, which contains glass fiber and mineral reinforcements, has excellent stiffness, a crucial factor to assure firm door sealing. Molded-in screw holes in the appliqué eliminate the need for drilling and allow precision fitting to the space frame.

5.1.7 Carbon Fiber Concept Cars

Carbon fiber is appearing on concept cars at both ends of the market. While the $15/lb price of carbon fiber versus a market requirement of $5/lb limits wide acceptance, its high strength, low weight, and prestige make the material attractive. In terms of prestige, at the high end of the market, Chrysler's concept ME Four-Twelve sports car, which if produced would sell for over $100,000, has a carbon fiber/aluminum body, and carbon fiber is also used for the seat structure frame. At the other end of the market, Chrysler is looking at the material's lightweight to squeeze more than 45 mpg in fuel usage from its Slingshot concept vehicle, a small city car based on the European Smart roadster. For now, however, the composite's high cost will not permit its use in a vehicle intended to sell for under $20,000.

While few concept cars may be commercially produced with carbon fiber, its use concept applications shows there is growing interest in the material. Many from composite experts, to universities and automakers are studying ways to make the material more economical. Though the raw material cost is dropping, there is as yet no way to mold carbon fiber in high enough volumes to production costs. Auto companies are currently making limited use of the material strategically placing it where it is needed. Meanwhile, companies like Meridian Automotive Systems, a key North American producer of sheet molded composites (SMC), are working to lower the composite's cost.

5.1.8 Metal/Plastic Antenna Modules

A top of the line car is now equipped with an average of ten antennae, and that number is expected to soon increase to as many as 16 as car manufacturers introduce systems for monitoring and automatically regulating the distance from the vehicle in front, and install warning sensors to avoid collisions when reversing. These numerous antennae have been generally spread over the entire bodyshell, presenting a challenge to designers who have to guarantee optimum reception.

A new antenna module developed as a joint project by Johnson Controls and Volvo demonstrates a way round the problem. The module accommodates all the antennae and relevant receivers (radio, TV, GPS) in one unit. Made of steel and Durethan BM 130 H2.0 polyamide supplied by Bayer MaterialScience, it is manufactured using the plastic/metal hybrid technology already established for production of vehicle front-ends.

The 1.8-kg antenna module is located under a plastic cover on the rear roof of the VOLVO XC90 SUV. The antennae are integrated into a film that forms part of the plastic cover. The various fixing elements and backing for the film, which extends across the width of the vehicle, are also made of polyamide. The receivers, such as the FM/AM, Digital Audio Broadcasting (DAB), and TV tuners, are screwed to the metal support of the hybrid construction obviating the need for a large number of cables.

5.1.9 High-Value Oil Valve Covers

North America's first high-volume thermoplastic valve cover was introduced on the Chrysler Town & Country, Dodge Caravan and Grand Caravan with 3.3 and 3.8 L V6 engines. The valve cover, manufactured by Bruss Sealing Systems from DuPont Minlon mineral reinforced nylon, reduces weight by more than 65% and cuts cost significantly compared to metal. The innovative thermoplastic valve cover also has an integrated air/oil separator, which significantly reduces the amount of oil pulled into the engine and an integrated positive crankcase ventilation (PCV) valve housing, which helps reduce evaporative emissions to reduce environmental implications.

A Chrysler, Bruss, and DuPont Engineering Polymers team ensured this program went from CAD drawing to commercial launch in less than 22 weeks ensuring introduction for the current vehicle model year. The valve cover is made of glass/mineral reinforced Minlon that has a balance of stiffness, strength, dimensional stability, and warpage resistance. The switch from metal to thermoplastic eliminated costly e-coating and several secondary machining steps. Runners and scrap from the injection molding manufacturing process are melt recycled.

Demand for long fiber reinforced thermoplastic (LFT) composites in automotive applications is continuing to grow, particularly in Europe. Unlike conventional short fiber reinforced materials, the longer fibers in molded parts mechanically interact with each other to form an internal glass fiber skeleton that limits anisotropic shrinkage and greatly reduces warpage. The composite materials with high stiffness and low weight are excellent metal replacements. The high level of long glass fibers also gives them metal-like coefficients of expansion. A nylon base material filled with 50% glass fiber has excellent resistance to automotive fluids and can tolerate the high-temperature exposures of this application.

SABIC produces Stamax P polypropylene LFT, a material with good thermal/ mechanical properties suitable for semistructural applications. It is used to mold front-end modules on the Porsche Cayenne, VW Touareg, and Mini Cooper. Meanwhile, LNP Engineering Plastics' Verton MFX, a 40% LFT PP is used to injection mold Mazda 6 front end and door modules.

Ticona's Celstran 30% LFT is used to produce door modules and a 50% filled nylon 6/6 Celstran is used in the grille opening retainer (GOR) of the Jaguar XJ. DuPont Rynite PET polyester resin is used to mold a GOR for the Lincoln Navigator and Ford Expedition. The material's high resistance to heat sag allows it to withstand Ford's e-coating curing process. Front end carriers of 40% glass fiber/PP are produced by the Faurecia Group in France. These components are produced with a Krauss-Maffei IMC machine, which compounds the formulation on a twinscrew extruder yoked to an injection molding machine.

5.1.10 Composite Front Hood

General Motors Chevrolet Corvette Z06 Commemorative Edition features a new engine hood using carbon fiber material. The carbon fiber hood represents the first time this advanced material has been used as original equipment for a painted exterior panel on a North American-production vehicle. The lightweight, race-inspired carbon fiber hood weighs just 20.5 pounds, 10.6 less than the standard hood, providing another measure of weight savings for a car that already enjoys a very potent power-to-weight ratio. Previously reserved for racing and exotic sports cars, carbon fiber combines extremely high strength and low weight. On most carbon fiber parts, the woven pattern of the material is easily seen beneath the exterior finish. To diminish that effect and the carbon fibers are aligned in a single direction. The overall result is a finish consistent with the rest of the car that gives just a slight hint at the carbon fiber construction.

The inside hood panel is a hybrid of carbon fiber and Sheet Molded Compound (SMC). The Class A outer skin supplied by Toray Composites (America), Inc., Tacoma, WA, is a carbon fiber/epoxy unidirectional prepreg, 1.2 mm (vs. SMC 2–2.5 mm) thick, comparable to stamped metal sheet and is the thinnest plastic composite skin on a production vehicle. The hood outer panel is molded and cured in an autoclave similar to typical aerospace composites. However, aerospace autoclave cycles often take 20 h whereas this Toray formulated prepreg cures completely in 10 min at 300°F.

5.1.11 Ford's GloCar

Ford Motor Co. is taking a lighter approach to design with its GloCar concept. The car uses light emitting diodes to change body panel color, and brightness in response to safety conditions and driver's whim. Launched last year, the concept car is on display at the Cooper-Hewitt, National Design Museum as part of its exhibit 'National Design Triennial: Inside Design Now.' The customizable GloCar was designed to be safe, fun, and evoke emotion. Illuminated by translucent body panels and LED lights, the soft glowing vehicle offers enhanced visibility. An additional safety feature stems from the fact that the car can be seen from all angles, not just headlights and taillights, making it very visible at night. The rear panel doubles as a brake light, and the side panels as blinkers. When somebody comes too close, the panels increase in intensity, signaling the driver to keep a distance.

The GloCar lighting combined with injection molding can potentially eliminate the need for vehicle paint, reducing manufacturing complexity. The concept car projects an image of concern, safety, intelligence, and lightness. Built as a design exercise it focuses on the prospective needs of drivers and shows a future where cars are more optimistic and intelligent. The vehicle, powered by fuel cell technology, is built around a lightweight aluminum space frame with aluminum extrusions and castings. There are no plans to produce the car.

5.2   Aerospace Applications

5.2.1   Aerospace Applications Introduction

During the past 50 years, aeronautics technology has soared, with plastics playing a major role in aircraft, missile, satellite, and shuttle development for civilian air travel, military air power, and space exploration. Plastic materials can be flexible enough to withstand helicopter vibration but rigid enough to ensure safe cargo. They can be transparent for easy observation, shatter resistant and offer ballistic protection, and, most significantly, they can be both lightweight and strong. Aerodynamic requirements of aerospace products demand maximum design flexibility and minimal weight. Plastics can be formulated to meet a wide variety of specifications and are ideal for components incorporating smooth curves. Composites are widely used in the panels of military jets and helicopters as well as for wing skins, nacelles, fairings, flaps and helicopter rotor-blades in commercial applications. Plastics are also found throughout aircraft interiors in bulkheads, galleys, stair units, seating, and flooring.

5.2.2   Unmanned Aerial Vehicles

As in conventional aircraft, minimizing weight is important in the new unmanned air vehicles (UAVs). Some, such as Northrop Grumman's Global Hawk, are designed for prolonged flights at high altitude. Cruising at extremely high altitudes, Global

Hawk can survey large geographic areas with pinpoint accuracy, to give military decision-makers the most current information about enemy location, resources, and personnel. Although Global Hawk's fuselage is aluminum, most of its other structures, including certain fuselage components, are composite. Low airframe weight means that half the 25,600 lb maximum takeoff weight is fuel, providing 34 h of flight. The UAV's wing, longer than that of a Boeing 737, comprises four I-beam spars and prepreg face sheets, with carbon/epoxy spars and envelope bonded together. Leading and trailing wing edges are Nomex aramid honeycomb cored sandwich laminates.

5.2.3   A380 Composite Floor Panels

M.C. Gill Corp delivered to aircraft manufacturer Airbus Industrie in Toulouse, France, the first set of a new type of composite high-performance floor panels for the A380 cockpit and two electronic equipment bays. The complete set consists of 67 parts fabricated from Gillfab 5509 panels, which were specifically designed for this application. These high-performance composite products are low smoke aircraft sandwich panels of a Kevlar honeycomb core, with facings made of phenolic resin reinforced with cross-plied unidirectional carbon and a thin fiberglass overlay to protect against galvanic corrosion.

The raw stock panels were developed and manufactured at the M.C. Gill Corporation headquarters facility in El Monte, CA, then detailed at the M.C. Gill Europe plant, which has extensive knowledge and expertise in detailing complex floor panels for a range of aircraft and customers. Detailing activities included drilling, machining, folding, and fabricating the panels into precise configurations to meet engineering specifications. The equipment bay panels were treated with a non-slip surface.

5.2.4   Ultra Lightweight Global Flyer

Virgin Atlantic has unveiled the GlobalFlyer V. The pioneering aircraft, said to be the world's most efficient jet plane, achieved the first solo, non-stop around the world flight. If successful, the aircraft will complete the 80 h voyage on one tank of fuel. Virgin Atlantic is the sponsor the 38.7 foot long, single engine jet, with a 114-foot wingspan. Scaled Composites spent more than 4 years designing and building the aircraft. The company used computer aided aerodynamic studies and built the structure entirely from ultra-light carbon, Kevlar, glass, and a combination of reinforced epoxy and polyester composite materials.

On takeoff, the plane will weigh 22,000 lbs, with over 84% of that weight being fuel and less than 80 h and 23,000 miles later, it will have shed 18,000 lbs to land at its dry weight of a little under 4000 lbs. The GlobalFlyer's 3-body configuration (twin-boom fuselage, and seven foot cockpit pod) spreads the aircraft's weight across the wing reducing the peak bending load and aircraft weight. The flight will take off from the US, in early 2005, flying at 45,000 ft and traveling 21,300 miles (with jet stream) at speeds in excess of 985 mph. On completion of the trip, GlobalFlyer will have traveled 75% further than the range record set for jet powered planes.

5.2.5   Mars Tested Circuit Materials

Since the start of manned space flight, DuPont has been providing high tech materials, essential for lighter weight, reduced volume, durability, and environmental resistance. Dupont Electronic Technologies have a broad range of advanced circuit materials, many Mars tested on the recent rovers, Spirit and Opportunity.

Seventy yards of flexible cable circuits made of thin DuPont Pyralux laminates/ composites replaced bulky round cables to save 60–70% volume in each Mars rover. The Pyralux flexible circuits connected each rover 'brain' to their robotic arm, cameras, high gain antenna, wheels, and sensors. Pressure sensitive tape made from DuPont Kapton polyimide film was used throughout to control vibration. Pyralux flexible cables secured with Kapton tape offered durable, lightweight environmental resistance at Mars temperatures of –120°C to 22°C. Hundreds of Kapton strip heaters were used to maintain minimum temperatures in the extremely cold martian atmosphere and allow the rovers to use smaller solar panels and batteries. Kapton film worked with layers of DuPont Teflon fluoropolymer resin and Pyralux flexible cables to provide power from the Rover electronics module to hardware components. The flexible joints of the robotic arm, which had to withstand repeated bending in extreme environments, used Kapton and Pyralux. Kapton strip heaters were also wrapped around motors on the robotic arm to keep them running at optimum temperature and efficiency. Metallized Kapton was used in thermal shielding for heat-sensitive components. The airbags used for the rovers landings were threaded and reinforced with DuPont Kevlar fiber.

5.2.6   Robotic Jet Bombers

The same plastic foam technology designed for manufacturing surfboards is being applied to the structural core material for wings of the X-45A unmanned combat aerial vehicle (UCAV). Central to the Foam Matrix' technology is a unique, patented process for producing net molded structures without the need for multi-piece assemblies. Foam Matrix' highly integrated structures replace the large number of individual parts used in 'traditional' aircraft manufacturing with a single component that satisfies the requirements. Using the firm's 'Foam Matrix Core' (FMC) system for the X-45A, the ribs, stringers, electrical conduit. and other wing parts are machined into a single mold tool. The entire wing is molded as a one-piece foam core. On curing, the foam core is wrapped in composite fibers and returned to the FMC mold tool for resin injection and final cure.

The FMC system reduces parts count, provides seamless construction, reproducibility, shorter production cycle time. and lower cost. Use of the foam also eliminates structure hollow cavities, improves damage resistance in handling/assembly, and reduces the possibility of delamination. As the skins are backed by the foam, field repairs are easier. Boeing found the high-strength lightweight concept provided an attractive alternative to conventional wing designs. Phantom Works, Boeing's R&D unit, has built several X-45A UCAVs under a $191 million agreement with the Defense Advanced Research Projects Agency and the US Air Force. The wings are designed to be removed from the fuselage, and the disassembled aircraft can be stored for up to 10 years.

5.2.7   Parachute Safe Air Transport

Eclipse Aviation's new six-seater air taxis may boast a safety feature no jetliner can match: a parachute large enough to float the plane to the ground in the event of an in-air emergency. Eclipse is creating the six-seater low cost/luxury air taxis to transform business travel. Ballistic Recovery Systems (BRS) has a $600,000 SBIR (Small Business Innovative Research) award from NASA to develop a parachute for six-seat jet air taxis.

Originally designed for lighter weight ultralights and experimental aircraft, BRS has achieved the technological ability to develop parachute systems for conventional aircraft. The jet taxi parachute will be a scaled up version of the chute BRS already supplies to Cirrus Design for its four-seater propeller planes. BRS' unique proprietary emergency ballistic parachute system is deployed with a solid fuel rocket motor. The system, which makes use of extremely lightweight nylon/polyester fabric and components, is pressure packed under 40,000 pounds of pressure. The resulting package can weigh as little as 15 pounds (ultralight), 45 pounds (Cessna 150), or 67 pounds (Cessna 172). Chute system attaches to the airframe at three locations, two at the leading edge where the leading edge meets the fuselage and one aft of the baggage area. Cost of a typical system is generally 5–10% of the value of the aircraft. Avemco, the leading aviation insurance underwriter, grants a discount on insurance premiums for those pilots choosing to fly with the BRS system.

5.2.8   7E7 Dreamliner Composites

Boeing is introducing a series of innovative features with its new 7E7 'Dreamliner' airplane, from initial concept and name to design, interiors, materials of construction, and final assembly processes. The plane, recently introduced to customers, is expected to be available for service in 2008. The new super efficient jetliner will be the first commercial jetliner to have the majority of its primary structure, including wing and fuselage, to be made of composites rather than aluminum. The redesigned interior will feature dynamic lighting, larger lavatories, more spacious luggage bins, and electronic window shades with adjustable transparency.

The company conducted extensive analysis before choosing, a carbon/epoxy composite for the primary structure and a titanium/graphite composite for the wings. Composites offered numerous advantages including better durability, reduced maintenance requirements, and increased potential for future developments. Boeing believes the choice will help them to take advantage of the most modern materials technologies as it starts into the second century of flight. Toray Industries' Torayca prepreg composites will be used in the primary structural areas. This material, a combination of high-strength carbon fiber and toughened epoxy resin, has been used on the Boeing 777 in structural applications such as the empennage and floor beams. Projected to be 20% more fuel efficient than current jets, the 7E7 and 7E7 Stretch will carry 200–250 passengers on routes of up to 6,600–8,000 nautical miles.

5.3   Marine and Boat Applications

5.3.1   Marine and Boat Applications Introduction

Because they provide superior weight to strength ratios and resist weather and salt-water much better than metal or wood, plastics have found numerous applications in the marine industry. The real plastics revolution in boatbuilding began in the 1940s, when fiberglass hulls made their debut. Boats were sleeker, better looking, more fuel efficient, more maneuverable, and easier to maintain. Today, boatbuilders looking for strength and lighter weight turn to plastics for sails, rudders, dagger boards, centerboards, slats, spars and wings, as well as hulls. Composite use in the nautical world gained considerable notice when America's Cup entries began using advanced composites in futuristic designs for both hulls and masts. Today, composites and other plastics find wide application both on and below the sea in applications such as personal watercraft, jet boats, sail boats, stern drive boats, outboard boats and inboard boats, diving equipment, off shore platforms, as well as commercial and military vessels.

5.3.2   State-of-the-Art Outboard Engines

After investing 5 years and $100 million in its new line of large, 4-stroke outboard engines, Mercury Marine's sleek new 'Verado' outboard engine cover is a real winner. Besides visual appeal, the engine includes some technical innovations. In comparing compression molded SMC, thermoforming, RIM and injection molding, the design team found that injection molding would afford greater design flexibility, parts integration, lower costs for assembly and paint, lower overall weight, and higher yields.

The upper cowl is the largest injection molded, glass reinforced nylon part ever made. It weighs 11 lbs, 20% less than if made from Mercury's traditional material, compression molded SMC. The combined upper/lower nylon cowls weigh 30% less than with SMC and cost 46% less than SMC models. The engine cover has to be structurally sound, able to withstand hitting a floating log while traveling at 40 mph. The design team spent 18 months to come up with an aggressive, powerful design for the Verado. Injection molding allowed additional design freedom, giving crisper edges and a high gloss look on some of the parts out of the mold so they would not have to be painted. The final cowling design consists of four primary parts, the top cowl, rear cowl, a structural rib, and front cowl, along with two mating parts that comprise the lower cowl. Glass filled nylon 6/6 is used for the top, rear, and rib members, whilst the front is molded from mineral/glass filled nylon. The massive 635 lb, 4-stroke, 275 horsepower (200, 225, 250 hp also available) engine, Mercury's first supercharged outboard for the consumer market, will retail for about $18,000.

5.3.3 High Tech Pontoon Boats

Genmar Holdings, Inc. introduced its latest technological innovation, the Windsor Craft pontoon, at the Minneapolis Boat Show. The pontoon market segment has grown more than 50% during the past 5 years. Genmar, the world's largest builder of recreational boats, is producing the mostly rotational molded 24 ft tri-hull pontoon from more than 2,600 lbs of polyethylene. Unlike a traditional pontoon's aluminum hull, the PE withstands saltwater, alleviating concerns about saltwater corrosion. The tri-hull pontoon boat, constructed using the company's patented Roplene construction process, features a lifetime, limited warranty on the hull and enhanced performance characteristics providing big-water handling while offering a level of comfort and style previously unavailable in a pontoon boat.

Roplene is a patented dual-wall construction system made by rotomolding marine-grade PE. This advanced technology was developed by Genmar subsidiary Triumph Boats and allows Genmar to build higher quality, stronger boats with limited lifetime warranties at a lower cost to the boating consumer. LLDPE is used for the three hulls, fold-up cabana/changing station, console and base structure; high-density PE is used for the floor and sides; and cross-linked PE for the fuel tank.

5.3.4 Subsea High-Strength Moorings

Single Buoy Moorings of Monaco is using PTFE filled Victrex PEEK polymer to overcome corrosion problems with bronze bearing and thrust washer components used in the company's subsea mooring systems. The company is a major supplier of floating production, storage, and offloading systems (FPSOs). Designed and molded by 3P/EGC of Texas, the polymer bearings and thrust washers in the mooring and driving chain systems have an extended service life, significantly less weight, and excellent load bearing capabilities compared to those of the bronze parts.

Corrosion resistance in subsea environments is of particular importance since these platforms are constructed in open or unprotected water to facilitate the mooring of tankers at offshore oil/gas exploration stations that have no jetties and breakwaters. Unlike bronze components, which can develop severe galvanic corrosion, PEEK bearings and thrust washers provide long-term corrosion resistance and superior wear resistance. The bronze parts are subject to premature wear and reduced service life should they run dry unlike PEEK parts, which require no external lubrication. The PEEK material also provides significant weight savings, important as the mooring bearings and thrust washers can be up to 50 cm in diameter, with wall thicknesses of up to 5 cm, and are extremely heavy and difficult to handle when they are made of bronze. 3P/EGC has developed a tailor-made process to mold the large PEEK polymer cylinders without defects.

5.3.5   Small Craft Rotomolding

Ideally suited to the rotational molding process, rotationally molded kayaks, essentially hollow parts (with a hole cut for the person to enter), have a strong position in the small marine craft market. However, fiber reinforced plastic kayaks also maintain a good market position. Competition is similarly vigorous between rotomolded and FRP canoes as well as other small boats up to about 20 feet long. Rotomolded boats are able to challenge FRP primarily through price with a rotational molded kayak selling for roughly 60% of the price of an FRP kayak.

Rotomolding is easier to automate and, while cycle times are long, they are measurably less than FRP cure times. Polyethylene raw material costs are also less than those of FRP, and PE parts have higher impact toughness, a strong selling point for small rotomolded watercraft often used in rugged environments. On the other hand, small FRP watercraft have a better surface finish/color and are capable of more innovative designs. Repair is also problematic for the rotomolded PE boats, requiring non-aesthetic thermally welded patches. Additionally, FRP parts can be made stiffer allowing lower wall thickness and all important lower weight. To increase stiffness, PE watercraft have recently been successfully rotomolded with a double hull and foamed interlayer.

5.3.6   Corrosion-Resistant Pier Sleeves

Many of New York's Hudson River piers, built with wooden piles, are more than 50 years old, have become corroded, and are losing their structural integrity. Additionally, with significant improvements in the river's water quality, wood boring worms, which eat through the wooden piles, are thriving in the river.

These corroded wooden pier supports are being restored using more than 1½ miles of fiber reinforced plastic (FRP) sleeving. MFG Construction Products Co., a leading supplier of composite products to the construction industry, produced the sleeves needed for the restoration using a general purpose polyester resin, reinforced with 30% glass fiber chopped strand mat. The sleeves fit around each pile to form a jacket, and epoxy grout is then pumped into the space between the jacket and the pile. Made without a gel coat, the sleeves are translucent, allowing grout levels to be monitored during installation. The pile repair sleeves are manufactured individually according to pile length and diameter. A diver wraps the sleeve around each pile then fastens it in place with a tongue and groove joint. Nylon straps are added to provide extra support and the grout is then pumped in. One diver can install 6–10 sleevesper day. Repairing the piles using the fiber reinforced plastic sleeves is much more economical than replacing the pilings or rebuilding the pier.

5.3.7   Lightweight Engineered Dock Pilings

Startup company ArmorDock Systems Inc., is using PVC to make pilings, boatlifts, fish cleaning stations, and other similar marine products. The new company is out-sourcing production to pipe extruder Hawk Plastics Corp. and Macon Plastics Inc. of Macon, GA, which handles the fabrication. ArmorDock chose PVC for its long 50 year service life, compared with timber's projected 15-year life in marine service. The PVC products are also light and easy to handle. ArmorDock was created in response to the need to find alternatives to timber treated with chromated copper arsenate (CCA). It is expected that the Environmental Protection Agency will phase out CCA-treated timber for marine and water applications, following the phase out of wood treatment in land applications. The PVC pilings, available in 8–12 in. diameter, do not require reinforcement, but can be filled with a substance such as crushed stone. The firm has four patents and is working on other products including sea walls and bulkheads. In pile accessories are also produced, including in-pile dock lights, in-pile water stations, in-pile electrical receptacles, and in-pile charging stations. ArmorDock PVC is impact modified, so it can withstand abuse applied when it is driven into the ground and can also withstand the impact of boats.

Chapter 6
Consumer Products End Use Applications

Keywords Consumer products • Polycarbonate • Blu-ray • High temperature • Toy • Recreation • Leisure • Furniture • Lawn • Garden

6.1 Consumer Products Applications

6.1.1 Consumer Products and the Consumer

Consumer products cover a lot of territory – everything from cosmetics and toothbrushes to housewares, toys, recreational goods, furniture, lawn and garden equipment, etc. Considering the diversification of this category, it is extremely difficult to assign overall growth figures though there are a number of very profitable niche opportunities. Trends in disposable personal income play a major role in the purchase of these products with any growth in real wages boosting consumer spending in this sector. The downside of business in consumer goods is that the product life cycles are often short, and most applications are price sensitive. Also, many of the end uses have seasonal patterns.

6.1.2 A New Wine Cork Twist

Solvay Advanced Polymers' Amodel Polyphthalamide (PPA) has broken into the wine industry. Typically found in automotive, electrical, and industrial product applications, Solvay Advanced Polymers LLC's high-performance polymer has found a new use as a threaded anchor for the patented twist-to-uncork wine packaging designed by Gardner Technologies, Inc., a Napa, CA-based manufacturer. The new product carries the MetaCork trade name and was on display at the recent K Fair in Düsseldorf. MetaCork™ consists of a hard plastic capsule with a threaded interior

D.V. Rosato, *Plastics End Use Applications*, SpringerBriefs in Materials,
DOI 10.1007/978-1-4614-0245-9_6, © Springer Science+Business Media, LLC 2011

surface, a matching plastic threaded cap, and a natural or synthetic cork fitted with a threaded anchor, made from Amodel PPA, that is screwed into the cork during the bottling process.

Twisting the plastic capsule eases the cork out of the bottle, thanks to mated threads inside the capsule and on the neck of the bottle. Once removed, the cork along with the anchor and the top cap can be pushed out of the capsule unit. The plastic capsule can be returned to the bottle for drip-resistant pouring, and the cap can be screwed back on for a leak-proof seal. The design of the anchor is critical to the design of MetaCork since the average breakaway torque required between the anchor and capsule to extract a natural cork is about 14.5 inch-pounds. Gardner Technologies selected Amodel PPA because it can withstand more than twice that level of torque. The Amodel material used in the MetaCork is a 45% glass-reinforced grade of PPA resin. The material has very good flow characteristics, making it easy to mold the threaded anchor or other parts with complicated geometries.

6.1.3 High Tech Gloves

Ergodyne, a pioneer in the development of products for worker comfort and safety, has introduced ProFlex X-Factor gloves, designed to offer a new standard in protection for extreme work environments such as those encountered by emergency, rescue and construction workers. The specialized gloves offer unique design elements to deliver the required combination of protection and dexterity. Features and benefits of the ProFlex X-Factor gloves include Keprotec, which employs DuPont's Kevlar (730/732) or Armortex, which employs DuPont's Kevlar/Nomex (726/728) over Amara synthetic leather on fingers, thumb, and palm to reinforce against abrasions. Double-needle Kevlar (730/732) or double-needle 3-ply 280D nylon (726/728) stitching provides added durability. EVA padding protects knuckles against scrapes, cuts, and bruises and also dampens shock, impact, and vibration. The use of reflective nylon on the glove backs improves worker visibility. The elastic cuff prevents the introduction of dangerous or uncomfortable debris. PVC fingertips prevent slipping and improve grip. A stretch spandex shell ensures maximum comfort and fit.

6.1.4 Consumer Durable Cushioning

Rogers Corp. offers a collection of high-performance foams for use in the communications, transportation, and consumer markets. The company's Poron urethane and Rogers polyolefin foams offer a broad range of design solutions for energy absorption in applications from insoles and sock liners to life jacket cushioning and back packs, where performance and comfort are essential.

Poron urethanes are a family of high-performance cellular foams that are ideally suited to footwear applications. Poron comes in a wide variety of colors, thicknesses, densities, and surface textures to accommodate footwear for any type of activity, from high-impact sports to elegant or formal occasions. A lightweight cellular material, Poron urethane provides insoles with greater shock absorption than many other materials. It also provides breathability, forming permeable layers that draw water vapor and perspiration away from the foot to help keep shoes and feet dry. Depending on the application, Poron urethanes combine readily with other materials, including fabrics, typical EVAs, and molded urethanes. It is also easily die cut. In all applications, it promotes durability, abrasion resistance, friction, tension, and stitch stress. Since Poron materials are made without plasticizers, the shoe insoles do not shrink or become brittle and crack with age. Rogers polyolefin foams are cross-linked elastomeric and plastomeric polyethylene-based foam materials offered in roll or bun form in a wide range of thicknesses, densities, and colors. The lightweight foams are thermoformable and easily heat laminated to other materials. They possess excellent tensile/elongation properties and are used in many sports and leisure applications.

6.1.5   High-Capacity Blu-Ray DVDs

Sony and arch rival Toshiba were on a collision course to work with film producers to make the next generation of high definition DVDs. Warner Bros., Paramount, and Universal have signed nonexclusive agreements with the HD Group, a consortium headed by Toshiba, to make the new HD-DVD format which was discontinued in 2008. Meanwhile, Disney signed a nonexclusive agreement with Sony to produce its videos as Blu-rays. Equipment makers and replicators monitored the situation closely in order to efficiently adapt the format the marketplace would ultimately choose. The battle was reminiscent of the 1980s fight between Sony and JVC over the creation of the first-generation of video cassettes. Sony opted for Betamax though the industry eventually adopted the rival VHS standard.

The Blu-ray disk, which Sony developed with Matsushita, holds up to five times the content possible with current DVDs, 25 GB for a single-layer BD and 50 GB for a dual layer, while a standard DVD can hold 4.7 GB of content. This technology is being pushed by Sony and a large coalition that includes most of the industry's largest consumer electronics and computer makers. The rival HD-DVD format, which was developed by Toshiba and was marketed along with NEC and Samsung, was expected to be on the market more quickly because it required fewer modifications to existing DVD technology. It had three times the current space on a DVD but was cheaper to produce than Blu-ray. The three studios that signed with Toshiba said they expecedt to begin selling some DVDs on the Toshiba-backed format. HD-DVD machines would be able to play older discs and also would enable replicators to retool without major equipment changes for the HD capability. Replicators of the revolutionary Blu-ray format require mostly new equipment. Sony's new device records and plays back Blu-ray discs but can also play back DVD, DVD-RW, DVD-R, CD, and CD-RW discs. However, the recorder will not be able to read DVD-RAM or DVD+RW discs.

Resin supplier Sabic Innovative Plastics had a stake in the contest. The company explored qualifications of possible Blu-ray materials over 2 years. Noryl EXLN0090 has proven itself in Blu-ray trials vs. optical-grade Lexan polycarbonate. The Noryl product "shows better dimensional stability than standard PC." BD materials will be more expensive but a BD system requires one molding machine instead of two and one sputtering machine instead of two.

6.1.6   Electronic Book Binding

Sagoma Plastics invests in its own ideas, investing 10% of sales in product and production development. One of Sagoma Plastics' latest, patent pending, proprietary products is an attractive, high-quality package for optical disks that combines the familiar look and feel of a traditional book with the versatility, durability, and high quality of injection molded plastic pages. Sagoma calls it DigitalBook. Disks are placed in either a single- or double-sided tray that can be leafed through like pages in a book. A single DigitalBook can hold up to 34 CD or DVD discs. It uses custom-colored disk-holding 'pages' molded from ABS, PC, or HIPS, materials that are stronger than the crystal PS used in conventional jewel boxes. The DigitalBook design provides ample room to package promotional products, premiums, and literature along with the disks. Its multiple cover options include a classic hard book cover or soft cover treatment done in linen, leather, or paper, or a plastic pocket cover treatment where graphics are slid behind a clear plastic sleeve.

The market for the DVD/CD book packaging is potentially very large. Sagoma has patents on the high-speed automation to produce these products. They are investing $1 million for equipment and are looking for a larger manufacturing facility. The company expects to double their TCBS business over the next 2 years.

6.1.7   EU Food Compliant Containers

Rubbermaid has adopted a new EU food compliant grade of Santoprene thermoplastic vulcanizate (TPV) from Advanced Elastomer Systems (AES) for the seal in its food container lids. Following European food contact legislation, AES developed a new TPV that meets the latest legislation and is ideal for applications in food and beverage markets. AES collaborated with Rubbermaid to ensure that the new EU food compliant TPV grades would match Rubbermaid's part design and processing requirements. Complying with the latest requirements of the Plastics Directive 2010/72/EC for aqueous-based food, the new grades exhibit good chemical resistance, outstanding compression set and temperature resistance with excellent sealing properties across a wide range of temperatures. As a result, they are microwave, freezer, and dishwasher safe. The new Santoprene TPV in grades are available in three hardnesses: Shore 55A, 64A, and 73A. Typical applications for these grades include food and beverage closure cap liners, beverage packaging applications, other food packaging applications requiring high performance such as sealability and high-temperature resistance.

6.1.8 High-Temperature Cookware

Tupperware, one of the world's leading manufacturers of plastic household products, has developed a range of new tableware, in cooperation with BASF. The new attractively designed tableware is lightweight, durable, and also capable of withstanding temperature extremes. To meet Tupperware needs, BASF researchers developed a novel modified Ultrason product. The new material, designed for the Tupperware product, is specially formulated polyethersulfone (Ultrason E, PES), to which BASF added UV stabilizers. The material is break and impact resistant and particularly tough. This thermally stable thermoplastic can withstand temperatures from −50°C to 220°C without effect. The innovative plastic tableware can therefore be transferred directly from deep freezer to microwave and then placed on the table as attractively designed serving dishes. The material, suitable for food contact applications, has excellent resistance to food staining and dishwasher detergents. Because of its excellent performance, Tupperware is giving a 30-year guarantee on the new tableware. The transparent containers that reveal the contents at a glance are available in a variety of colors. Other possible applications are baby bottles and food service applications for airlines, restaurants, and large-scale catering kitchens.

6.1.9 High Tech Power Shavers

Gillette's next generation triple-blade shaving system the Gillette M3Power is the world's most technologically advanced and expensive mass-market wet razor. The battery powered plastic laden upgrade to the Gillette flagship Mach3 line is a unique razor. Users press a button on the handle to activate a small motor to produce a gentle pulsing action said to stimulate hair upward and away from the skin to produce a closer shave. The M3Power, which is said to outperform all men's shaving systems, including Gillette's Mach3Turbo, operates under 62 patents. Other innovations in the shaver include proprietary blades enhanced by a new coating, called 'thin uniform telomer,' which provides a perceptible improvement in shaving comfort throughout the life of the blade. The blade cartridge features an Indicator Lubrastrip infused with Vitamin E and Aloe for added moisture, and an ergonomically engineered handle with strategically placed gripping surfaces to enable one to shave safely at any angle. M3Power also is shower-safe, allowing convenient shower shaving. The shaving system includes the refillable razor, two cartridges, and an AAA Duracell battery.

6.1.10 Blow Molded Standup Tubes

Graham Packaging, a worldwide leader in the design, manufacture, and sale of customized blow-molded plastic containers for the branded food/beverage, household/personal care, and automotive lubricants markets, has developed a new packaging phenomenon. The unique design is a blow-molded stand-up plastic tube, which when squeezed exposes a pop up dome revealing a small, 13 mm diameter threaded cap from a recess in the tubes base. The small resealable closure and dome can be retracted back into the tube base when not in use so that the container can stand on end. In order for conventional tubes to stand upright, a large and costly cap is required to serve as a base on which to stand.

The patent-pending Graham Flexa Tube is constructed in one piece with no welds. Simple and sturdy, the new tube design is easier and less expensive to manufacture than are conventional tubes. The pop up dome has a living hinge that allows the dome to pop in and out of the tube base. The sealed end of the tube can be formed with a hook or eyelet, so that it can be hung from a display rack. The tube is available in 40, 50, and 58 mm diameter sizes. Made with LDPE, the packaging was initially introduced as a hair care product container. The package is being considered as a container for a range of products from creams and lotions to spreadable foods, sauces, condiments, and household/chemical products.

6.1.11 Do-It-Yourself Faucets

Moen Inc., one of the world's largest residential and commercial plumbing products producers, is manufacturing new kitchen faucet deck plates to make it simpler for plumbers and 'do-it-yourself' consumers to install. The plates have a built-in thermoplastic vulcanizate (TPV) lip seal of Santoprene TPV from Advanced Elastomer Systems, LP that creates the watertight barrier between the faucet and the countertop, eliminating the need to apply plumber's caulk during installation to achieve the seal. The pre-installed TPV gasket makes installation faster, and reduces tools and materials necessary for faucet installation. Installing a Moen kitchen faucet with the redesigned TPV deck plate seals simply requires inserting the faucet through the sink top, tightening the nuts, and attaching the watertight supply lines. The innovative new deck plates will be included in all of the company's kitchen faucets.

During a two-shot injection molding process, a narrow groove around the perimeter of the oval deck plate is injected with a nylon bondable grade of Santoprene TPV to create the lip seal. The TPV forms a virtually unbreakable bond to the durable nylon deck plate to produce a watertight surface. This process enabled Moen to reduce cycle times versus a previously used procedure that injected an EPDM/nitrogen-filled foam into a groove to create the gasket.

6.1.12 Time Expired DVDs

Sabic Innovative Plastics, in a 2-year development program with Flexplay Technologies, has helped develop a new patented Lexan polycarbonate copolymer that sets a precise limit on the service life of DVDs once they have been removed from their package and exposed to air. A Flexplay enabled DVD is similar to a conventional DVD, except that it has a 48-h viewing window that begins when the disc is removed from its packaging. Consumers will then be able to enjoy the movie or other product as many times as they wish during this time frame. After 48 h, the DVD will no longer be readable by the DVD player and can then be recycled. A Flexplay enabled DVD works in all players that accept a standard DVD. The secret to Flexplay discs lies in the extra polymer layer added to a standard DVD structure. This specialized layer includes a chemical compound that readily combines with oxygen. While the orginal chemical compound is transparent, the oxidized compound is opaque. When the layer is transparent, the laser passes through to the reflective layers to read the DVD data as normal. When the extra layer becomes opaque, data on the DVD can no longer be read. The polycarbonate disk also changes color from red to black to indicate that the disk service life has expired.

6.2 Toy Applications

6.2.1 Toy Applications Market

Retailers are devoting less shelf space to traditional toys and more to highly profitable video games. Larger, higher-priced toys such as ride-on toys and playground equipment (typically rotomolded) that offer lower sales for the amount of aisle space they consume as a result are feeling the crunch. Many small companies participate in the toy market, but four major injection molders dominate. They are Fisher-Price (part of Mattel), Mega Bloks, Little Tikes Co. (part of Newell Rubbermaid), and American Plastic Toys. Together they represent 77% of the injection molded toy business in North America. Polypropylene is the most common material utilized in this market, representing approximately 38% of injection molded toys. Second biggest is HDPE at 25% of the market. Polystyrene, ABS, polycarbonate, and LDPE resins are also utilized in considerable amounts, but each constitutes less than 10% of the business. Fastest growth (3.8%/year) is forecast for PS; next are PP at 3.2%/year and HDPE at 2.3%/year. All the other resins are expected to decline in poundage from 2010 to 2012.

6.2.2   Rotomolded Toys

Step2 Co., a multinational manufacturer and marketer of high-quality plastic products for children, makes extensive use of rotational molding, a process that Step2 founder and former president of The Little Tikes Company Tom Murdough pioneered for use in the consumer products market.

While manufacturing toys using rotational molding allows for tremendous creativity, to successfully push the envelope and get the most out of the rotational molding process, it is important to understand how the molds are made. For example, designers can prevent warping on large surface areas, such as the sides of a children's climber, by adding subtle features like crowning, contours, textures, and a slight 'pillowed look.' Mass retailer concerns add another design dimension for rotomolding designers. At Step2, many of the products are big and bulky. The company expends a lot of effort designing for packaging and assembly to minimize product package size. One of Step2's latest toys, its 'Naturally Playful' picnic table, has fold out legs that the customer snaps into place for a permanent fit – a design that saves considerable space in the box.

6.2.3   Large-Scale Building Toy

Sagoma Inc. is one of the few toy producing companies located in the US. The company produces popular, competitively priced toys bearing its trademark that are marketed and sold by its strategic ally and customer, Taurus Toy Co. of Portland, ME. One successful toy is its 'No Ends,' a large-scale building toy that allows children to build large-scale forts, doll houses, puppet theaters, helicopters, airplanes, animals/dinosaurs, and more from basic building components. The toys encourage children to learn to build, create, work in groups, develop three-dimensional relationship skills, and have a lot of fun. Though children can build things to large scale, the parts are in relatively small boxes, less than $24" \times 24" \times 24"$.

The company also does proprietary product development for other toy companies, and helps design, develop, build Stereolithography model/prototype, and assists in the tool building for such products to make for optimum efficiencies in molding. The company, a state-of-the-art injection molder and assembler, provides both product design and/or injection molding services. The designers and engineers at Sagoma are responsible for both new product design and daily manufacturing. The company enjoys several alliances with local companies, rapid prototypers, machine shops, and tool shops, which have lasted more than 30 years.

6.2.4   Highly Precise Game Joy Sticks

Logitech's latest controller for the PlayStation2 is the Logitech Flight Force joystick. Joysticks are all about precise and natural control of flight and space combat games.

In addition to excellent control, Flight Force provides the exciting immersiveness of Logitech's force feedback technology. This force feedback joystick lets gamers experience the sensations of flight combat, including weapon recoil, missile launches, explosions, and other effects, in compatible games for PlayStation2. Soft touch TPV, ABS, and polycarbonate plastics are used throughout the application. Logitech Flight Force features a twist handle with versatile handgrip positions for easy rudder control and improved hand comfort in tense game situations. Seven action buttons, four on the handle, two on the base, and a trigger, provide a broad range of control options to deliver easy access to fire weapons, select targets, and perform other functions using both hands. The handle buttons and an eight-way hat switch are designed for quick access. A high precision throttle provides accurate power modulation, while the substantial base, which contains steel weights, ensures stability when the going gets exciting.

6.2.5   High Tech Learning Toys

Electronic learning aids are one of the few growth categories in the $25 billion US toy industry. Parents spent more than $820 million on electronic learning toys for preschoolers in 2010 alone, despite a lack of proof that they will jump start toddlers on the path to perfect SAT scores. A sizzling hot product is LeapFrog's LeapPad, a learning aid that emits sounds and music when kids touch a stylus to a picture or letter. In what is shaping into a blistering toy battle, Mattel's Fisher-Price responded with its similar smash hit, introducing its PowerTouch. LeapFrog countered with three new products and slapped its rival with a patent infringement lawsuit.

In trying to beat LeapFrog, Mattel endeavored to make its PowerTouch unique. While kids use a stylus pen to interact with a LeapPad, they only need to use their fingers to operate PowerTouch. In addition, the PowerTouch console automatically recognizes when a page is turned. LeapPad users must push a button to continue, something that can confuse small children. LeapFrog has since come out with Littletouch LeapPad, basically a touch-sensitive plastic slab. By placing a specially designed book in a holder on the slab, and inserting a program cartridge, the pictures on the book pages become touch-sensitive buttons. The products are possible as a result of high-speed chips and compact memory. The same capabilities a few years ago would have required a CD-ROM and a $2000 computer.

6.2.6   Import Proof Seasonal Toys

Unlike many plastics processing markets, foreign competition is not a big concern at seasonal products blow molder Grand Venture. Seasonal blow-molded products are big, bulky, and do not ship well, as they cannot be stacked and they contain a lot of air. The firm's top Christmas product was 38 in. light-up Santa. The seasonal products company also offers a range of blow-molded nativity scenes, stars, lambs reindeer, snowmen, and other Christmas subjects. The company's largest Christmas themed blow-molded product is a 68 in. light-up Santa in a reindeer-drawn sleigh that the company recently introduced.

All of Grand Venture's products are made of HDPE and are molded by its sister firm, Falcon Plastics Inc., which shares its Washington location. Falcon had sales of $12 million last year and also blow molds trash pails, and lawn and garden products for Grand Venture. Grand Venture typically takes seasonal orders for Halloween and Christmas in Q1 of the year. It began molding Santas and other Christmas products in July, to ship to its biggest customer Wal-Mart Stores Inc. Christmas is still the top market for seasonal plastic products marketer.

6.3   Recreation and Leisure Applications

6.3.1   Recreation and Leisure Applications Introduction

In the US, the wholesale value of manufacturers' shipments of sports equipment and recreation-related products climbed to $120 million and $334 million, respectively, in 2010. Similar to toy manufacturing, production of many sport and recreation products have moved off-shore to Asian molders. Also like many other market sectors, traditional materials such as canvas, leather, wood, and metal used in the manufacture of sports and recreational products are increasingly displaced by a wide range of plastic materials. Products such as hiking equipment, camping gear, climbing boots, backpacks, snowshoes, skates archery equipment, and many others, even kayaks, are increasingly made of high tech composites and other plastic materials primarily for their excellent strength to weight ratios, water resistance, abrasion resistance, etc.

6.3.2   Ultralight Foldable Kayak

Firstlight Foldable Kayaks of New Zealand have launched a new range of foldable, strong and exceptionally lightweight kayaks. These dismantable kayaks are made up of a flexible outer skin in Desmopan TPU and a frame in carbon fiber/Kevlar fiber. The skin for the kayak is produced by Epurex, a Bayer MaterialScience subsidiary, specializing in extruded Desmopan TPU film. The entire hull is made from a single piece of material, meaning there are no seams under the waterline. Desmopan (TPU) was chosen for its high abrasion resistance, flexibility over a wide temperature range, good resistance to weather exposure, transparency if necessary, freedom from plasticizers, so it does not shrink and its excellent chemical resistance. These advantages permit a dimensionally stable, 3D outer skin which is tear-resistant yet still lightweight.

The kayaks, weighing 8–10 kg, fold down into a backpack-sized pack. Assembling the kayak takes about 20 min. Use of composite carbon fiber and Kevlar rods is the key to providing the ultimate in lightweight folding kayaks. Not only are these rods light and strong, but they provide excellent torsional rigidity. The cross ribs made from Dupont's Zytel high-performance polyamide are exceptionally strong, yet will flex to allow the rods to conform to the shape of the kayak. The three-dimensional shape of the bow and stern is formed using lightweight molded foam end units that cushion and protect the inside of the skin from damage caused by chaffing from the frame and reinforce the bow shape.

6.3.3   Clear Corrosion-Resistant Pool Cleaners

Sabic Innovative Plastics polycarbonate/polyester alloys are increasingly being used for a variety of pool and spa applications. Xylex X8210 and X8300 are transparent, have good impact strength, weathering performance, and resistance to chemicals like spa water conditioners, bromine, and chlorine making these alloys suitable for cleaner housings, light lenses, spa remote controls, pump covers, designer caps, filter parts, control panels, and other pool and spa accessories.

Xylex alloys combine an aliphatic polyester with polycarbonate to offer an outstanding balance of chemical resistance, clarity, and mechanical strength. The materials' strong resistance to chemicals used in pool/spa care and other aqueous environments helps reduce cracking/crazing and their good UV performance, and temperature resistance provide weatherability that is as good or better than many competing resins. Both Xylex X8210 and X8300 combine clarity and ductility with chemical resistance, high in mold flow, hot and cold ambient temperature impact strength, gloss retention, and UV resistance. X8210 resin is used for a variety of applications demanding impact strength, ductility, and aesthetics differentiation using GE Plastic's Visualfx special color and effect resins, while X8300 is better suited for applications requiring the highest level of chemical resistance.

6.3.4   Cross Country Skiing/Skating

All-terrain skates from GateSkate, Inc. are designed to go over any type of surface including asphalt, dirt, gravel, grass, and mountain trails. Thanks to an innovative design with a rugged chassis molded from a new super strong DuPont Zytel HTN high performance polyamide resin, and pneumatic tires, TrailSkate rolls easily over bumps, cracks, and debris, while hand-operated hydraulic brakes allow skaters to easily control speed and stop quickly while remaining comfortably balanced.

The chassis, essentially a ski on wheels, must withstand repeated dynamic loading under the rider's full weight. Long glass fibers in the 50% glass-filled Zytel HTN51LG50HSL resin are crucial to the chassis' performance providing improved fatigue performance, impact resistance, and surface appearance. The design is a challenging one. The 21 in. long chassis has wall section thicknesses of 0.04–0.79 in. Load-bearing requirements make avoidance of stress concentration essential and therefore mold design proved critical. DuPont US specialists using mold flow analysis recommended the gate location, and local specialists in China recommended enlargement of the gate to enhance fiber orientation and improvements in venting to assure complete mold fill.

6.3.5 Snow Skates

Global sports equipment manufacturer Salomon is using DuPont Delrin acetal resin for the ski section of its new concept, 'bombproof' PP3 Kurb snow skate. PP3, the codename for a well-known, 'bombproof' Russian submarine, was selected to communicate that the PP3 KURB resists shocks and damage from small stones. Snow skating, a snowboard-type sport recently launched in the US, differs from snow boarding in that the user is not attached by bindings to the board, giving the snow skater a greater feeling of freedom.

The acetal resin chosen for the ski section of PP3 Kurb is ten times tougher than HDPE, the standard material for ski running bases, providing significantly improved mechanical resistance against small pebble damage. The acetal resin is also flexible while remaining stiff and strong. Users can flex the ski by pushing down on the rear of the maple wood deck to complete significantly more complex jumps and tricks than with previous snow skate models. The design freedom given by the acetal also means that Salomon could design side channels into the ski section for added stability when the user wants to ride over soft, mountain snow. Meanwhile, in order for users to use their snow skates in the evenings, down in the resorts, central 'grind channels' are included in the design for extra stability when the user is 'grinding' along handrail sections of skateboard parks.

6.3.6 Laserline Golf Tees

The Laserline tee, a golfing innovation, was developed by Europe's most successful golf player, Nick Faldo, using Bayer's Makrolon polycarbonate. The new tee promises precise ball contact, improved flight path, and increased hit quota which all translates into an improved game. Unlike traditional tees, the teepin has been moved 8 mm towards the flag and ensures the direct contact with the ball without touching the tee. The development improves ball contact at every tee off providing both amateurs and professionals with a more precise game. When developing the tee, the manufacturers, 'Spirit of Golf' and Nick Faldo, chose polycarbonate as the high tech material has exceptional shock resistance and is unbreakable, giving the Laserline tee a clear advantage compared to classical wooden and traditional plastic tees. The Laserline tee is completely recyclable and complies with European environmental requirements. The innovative tee is available in a range of colors and can be obtained in golf sports shops. Spirit of Golf, the exclusive Bally Golf licensee, recently introduced the Laserline tee to the United States market.

6.3.7   Light, Efficient Golf Club Merchandising Carts

Meese Orbitron Dunne (MOD) Co.'s American Rotational Molding (ARM) Group redesigned and manufactured a custom golf club merchandising cart for Callaway Golf Co. The merchandising cart, a key element in Callaway's custom club fitting system, is a portable scooter that holds 56 woods and irons and an IBM ThinkPad data acquisition system. It reportedly enables Callaway teaching professionals to gather data from each swing and recommend ideal club sizes with the aid of the company's proprietary fitting software.

MOD/ARM redesigned the cart to add strength, longevity, and visual appeal while guarding against the elements. A sleek, durable, rotationally molded polyethylene structure replaced the earlier fabricated metal and vacuum formed plastic panel structure. The new design replaced hundreds of parts and fasteners and associated assembly requirements with nine rotationally molded panels, three injection molded wheels, and one extruded tube to protect each club. The simplification increased production rates, aided quality assurance, and contributed to an overall cost reduction of 75% versus the original metal cart. While cutting costs, MOD/ARM engineers added numerous amenities and security precautions. The cart was redesigned to fit through standard-sized doorways. A cumbersome front wheel assembly was replaced with a single front wheel that swivels 360°. A new brake halts the cart when movement is not desired yet remains hidden from view for visual appeal. A clever towing bar, which stows inside a compartment at the front of the cart when not in use, was added to permit towing with a golf cart.

6.4   Furniture Applications

6.4.1   Furniture Applications Introduction

The furniture market consists of five primary categories, home and office furniture, juvenile or infant furniture, public seating (school, lecture hall, stadium, and arena), outdoor and patio furniture, and miscellaneous furniture parts. Injection molded furniture consumes the most plastic resin. In 2010, the furniture market accounted for 4%

of total thermoplastic sales. In the United States, domestic injection molded furniture consumed an estimated 409.9 million lb of resin in 2010. With an average annual growth rate around 3%, this market could grow to 475 million lb by 2012. Polypropylene constitutes around 80% of the resin consumed for injection molded furniture. HDPE is a distant second place at around 7%, and polystyrene is third at 6%.

Within the injection molded furniture category, outdoor furniture is the largest segment (44% of the total), home and office furniture is the second largest area (25%), juvenile and infant furniture represents 22% of the market, public seating is 6%, and the miscellaneous furniture segment represents 3%. Much of the injection molded furniture market is directly related to the housing market. Changes in the level of orders for housing-related furniture products arrive about 6 months after any sharp changes in housing starts.

6.4.2   Polyurethane Foam Quilting Replacement

Bayer MaterialScience AG has recently developed an extremely light polyurethane foam to replace the padded quilting that normally makes up about 10–15% of the volume of upholstered furniture. Known as HyperNova this highly elastic material, thanks to the spatial crosslinking of its pores, is more dimensionally stable than the polyester fiber fleece and other materials traditionally used in this service. Using HyperNova, the seat surface of upholstered furniture remains largely unaffected by frequent use, always recovering its shape when you stand up, leaving no wrinkles or creases and the upholstery padding will retain its resilience over many years. An additional benefit of this very soft polyurethane foam is its exceptional breathability: it is extremely permeable to air and can absorb up to 10% of its weight in moisture and emit it again later. The breathable foam was manufactured in close cooperation with Hennecke GmbH of Sankt Augustin, Germany. The environmentally friendly foam uses only carbon dioxide for blowing agent. For allergy sufferers, the new polyurethane foam is resistant to mite and microbe infestations.

6.4.3   Lightweight Composite Tables

Under the name 'Triple-E luxe tafels,' Dutch firm Tinga R&D BV is marketing lightweight, stackable tables that are far easier to handle than 'conventional' wooden furniture, thanks to a sandwich composite using Baypreg, a polyurethane system from Bayer. Sandwich composites were originally considered exotic materials primarily used in aerospace and more recently automotive manufacturing. Now thanks to a promising idea from Tinga R&D, they will soon be found in lecture or dining halls and cafeterias.

Setting up and taking down tables in a multipurpose room is usually a heavy chore. Tinga recognized the potential that sandwich composites with their low weight and high strength might offer the furniture industry. Using a sandwich design the Triple-E tables weigh only 12.5 kg despite measuring 80×200 cm. The core of these tables is made of a lightweight rigid PU foam, reinforced top and bottom with flax mats to absorb tensile forces. The external surface is coated with an extremely tough, washable plastic film available in imitation wood or colored finishes. The table edges are reinforced with solid wood. To manufacture a sandwich composite, the fiber mats are thoroughly impregnated on both sides with PU in a spray process. The sandwich components are then placed in a preheated mold and pressed into final shape at an elevated temperature (60–120°C) and pressure (6–8 bar). The PU cures in minutes. The sandwich composite requires no after treatment. This highly economical, one-step process produces a durable, very tough and rigid composite that can withstand even high mechanical loads.

6.4.4 Naturally Flexing Office Chairs

An attractive new chair, part of contract furnishings manufacturer Allsteel Inc.'s 'Get Set' line of office chairs, tables, and accessories, is its Get Set multi-purpose room seating/office side chair. The new chair, a comfortable, flexible solution for learning environments, is designed for both horizontal nesting and vertical stacking (up to four high) for versatile, easy transportation and storage. Though portability is a plus, the benefit of Get Set chairs is the hours of comfortable seating they provide. Unlike non-padded folding or stacking chairs often used in a learning setting, Get Set chairs focus on comfort, featuring seat cushions and a unique 'flex back,' which has a structure that flexes naturally, to counterbalance body weight and conform to support users of different shapes and sizes. An ergonomic sloping arm option also allows for natural movement of the body. Perforations in the chair back allow air circulation, an important consideration, as well. Durable multi-surface casters make movement easy.

Innovative Injection Technologies Inc. of West Des Moines, Iowa, injection molds several parts for the stacking office chair, from 18% glass-filled nylon, using gas-assisted molding.

6.4.5 Performance Enhanced Window Shades

DuPont Delrin acetal resin is used for the gears in an innovative new gearbox for Hunter Douglas's UltraGlide lifting system, which eases height adjustment of honeycomb window shades. The low-friction gearbox allows Hunter Douglas to extend their product offering to larger shades, by providing a mechanical advantage to lift weights up to 7 lb, about double that without a gearbox.

Performance Gear Systems, Inc. (www.performance-gear.com) designed the gearbox to meet Hunter Douglas's needs for smooth operation and low wear in a compact package that fits in the space limited confines of the shade's head rail. The gearbox contains three gears, input, parallel-mounted cluster, and output gear. Performance Gear Inc. injection molds all three gears from a specially formulated ultra low friction grade of Delrin containing 10% by weight of DuPont Teflon PTFE fluoropolymer in micropowder form. The connector is molded from general purpose grade Delrin 500P. Teflon's locked-in lubricating action ensures that there is no risk of fabric staining, a potential problem with externally lubricated resins. Other components of the gearbox are the housing, housing cap, and retainer. The retainer holds the gears in the housing and has an integral bearing shaft for the input and output gears. Performance Gear molds these parts from DuPont Zytel 132 F nylon resin, an unreinforced, fast cycling PA66 grade.

6.4.6 Renewable Resource Sleep Products

Pillows, comforters, mattress pads, and fiberbeds filled with NatureWorks fibers are beginning to appear on store shelves, offering consumers a new and natural-based alternative to traditional synthetic fibers. The new fibers made from an entirely corn derived product were developed by Pacific Coast Feather Co., a leader in the US and Canadian bedding industry. These new bedding products provide natural comfort and warmth with superior performance attributes versus polyester fiberfill. These new products provide a natural loft and resilience to give greater support while sleeping, while the insulating warmth and wicking attributes make for a dryer and more comfortable sleep experience. Overall, NatureWorks fibers combine the best physical characteristics of natural fills and polyester synthetics but with superior performance that will not diminish after washings or through extended use. The new natural-based fibers deliver great comfort, loft, and durability without the 'environmental shortcomings' of petroleum-based, polyester fibers. NatureWorks uses 20–50% less fossil resources in the production process than conventional plastics, resulting in significantly less CO_2 emissions.

6.4.7 Environmentally Friendly Chairs

The Mirra midlevel executive chair is Herman Miller's most advanced launch following their 'Design for the Environment' protocols, which provides guidelines that engineers/designers must meet at every stage For the Mirra that meant considering not just the types of resins used, but the breakdown of all materials including stabilizer in the nylon and pigment for the polypropylene. Working under the guidelines of the McDonough Braungart Design Chemistry system, HMI dropped the use of PVC, replacing it with nylon or TPE.

The effort affected the chair's overall design. In addition to comfort during its work life, the chair is designed for easy disassembly for recycling. A typical chair back support has structural steel with an overmolded plastic skin, and adjustment mechanisms of overmolded screws and springs. Nearly impossible to recycle, Mirra designers and engineers decided to use an all polypropylene seat back supported by a glass filled structural nylon spine, with a composite leaf spring. A nonfilled nylon webbing provides additional lumbar support. The chair can be disassembled for recycling within 30 s.

6.5 Lawn and Garden Applications

6.5.1 Lawn and Garden Introduction

The $30 billion US lawn and garden market which consists of three primary categories – outdoor equipment, supplies, and professional lawn care services – is expected to grow by 5.5% annually through 2012 as American homeowners, now at a record number, settle into their newly acquired properties and baby boomers enter the peak years for gardening popularity. Driving this market are extremely favorable demographics, continued homeowner affluence, and a new focus on push-marketing by the consolidating major players, a proliferation of lawn and garden retailers, and numerous new products introduced to satisfy the needs of millions of homeowners. Additionally, golf courses are big business for the lawn and garden market.

6.5.2 Novel Plastic Fencing

Kroy Building Products, Inc., a leading manufacturer of vinyl fencing, decks, railing, accessories, decorative structures, and specialty building products, has introduced a novel new vinyl fence technology that offers superior strength and performance. The new technology, named 'Fusion Fence,' uses sonic welding rather than glue during the assembly process, to create a stronger, more durable fence. Fusion welding the vinyl components to the main frame of the fence section produces a sturdier and more solid fence.

Kroy's new Fusion Fence also simplifies the purchase/installation of components, producing a single unit that is easy to purchase, transport, and install. Every Fusion Fence vinyl section is custom manufactured to precisely fit the customer's specifications. All Fusion Fence vinyl profiles are manufactured via an extrusion process, fabricated with computerized machinery and assembled with fusion vinyl welders for superior strength, durability, and aesthetics. The technology is available in a wide range of traditional and classic fence designs for both the new construction and remodeling markets.

6.5.3 Miracle-Gro Multipurpose Pails

Seeking to redesign its Miracle-Gro pail, Scotts Co., the world's largest marketer of branded consumer and garden products enlisted the aid of Unimark Plastics.

Unimark, a leading full-service supplier of innovative plastic and manufacturing solutions, specializes in high-volume, precision injection molding. Replacing a nondescript stock pail previously used to hold four different varieties of the water soluble plant food formula with an updated design has helped sales of the fertilizer to flourish.

Scotts wanted to improve its miracle gro market presentation from the stock container and involved Unimark right from the beginning of the redesign. Other Scotts goals for the new design were an easily removable lid and stackability. It also wanted to eliminate secondary printing on the lid and replace it with a molded in engraving of the Miracle-Gro logo and name. Unimark helped improve the polypropylene bucket through its Greenville-based Innovative Solutions design house, designing a domed lid that sheds water. To improve appearance, the edges of the container are rounded. A Miracle-Gro logo is molded on the lid. Another design change helps the buckets stack on top of each other like Legos, as each lid fits into the bottom of the pail on top of it. Unimark promotes pellet-to-pallet automation capabilities, providing cost-efficient solutions so that their customers can put extra money into the design and marketing. The finished Scotts product is a compact, sturdy and bright, high gloss garden container in a bouquet of four vibrant colors and a new packaging market for Unimark.

6.5.4 Non-Slip Garden Tools

Fiskars Corp., a leading manufacturer of scissors and knives, has introduced non-slip garden pruners. Fiskars requires any design to unify all the qualities of an ergonomic product, including quality, viability, safety, and comfort. Starting from this design philosophy, Fiskars looked to design three new garden fingerloop pruners. The company wanted to produce these tools with grips that were more comfortable, soft and slip-resistant than previous designs, making the pruners much easier to use. Fiskars initially considered using a modified grade of polyamide, the traditional material for grips in these applications, but this option proved to be too costly. Alternative materials were assessed, and it soon became clear that Santoprene thermoplastic vulcanizates (TPVs) would produce the desired level of comfort and softness.

The grips for the tools are produced using a two-shot injection molding process, consisting of a shot of black nylon 6 with 30% glass fiber (Durethan BKV 130) followed by a shot of gray Santoprene TPV 8291-70PA, eliminating the need for adhesives or assembly procedures.

6.5.5 Functional Stylized Glazing

Conservatory roof manufacturer Wendland is offering Sabic Innovative Plastics' new Lexan Thermoclear Easy Clean polycarbonate sheet in a product line to help their conservatory roofs to stay clean. The Sabic Innovative Plastics' Lexan Thermoclear Easy Clean sheet is part of the Lexan Thermoclear sheet range of lightweight, weather- and impact-resistant sheets extruded from Lexan resin. Lexan Thermoclear Easy Clean sheet features a special coating on the outside surface that provides self-cleaning properties.

Suitable for all applications that are exposed to rain or water, this specially treated Lexan sheet has a special hydrophobic coating developed by Sabic Innovative Plastics that reduces the angle of water drops on the sheet from 101° to 66°. As a result the drops of water get bigger and transport and drain off the dirt on the sheet. The multi-wall sheet has UV protection on both sides and the unique Easy Clean properties on the upper side. The sheet, which is produced in a range of thicknesses, offers high light transmission, good insulation properties, and good impact resistance. Sabic backs the self-cleaning properties of the product with a 3-year warranty, although tests indicate it will last significantly longer.

Chapter 7
Medical, Emerging, and Other End Use Applications

Keywords Medical • Catheter • Surgical • Emerging • PLA • Luminescent • Virtual layering • Government • Pipe • Waste management

7.1 Medical Applications

7.1.1 Medical Applications Introduction

Medical is a strong and growing market, particularly as the baby boomer generation enters old age. The industry projects demand for disposable medical supplies will grow 5.6% annually through 2012. Gains will be led by drug-eluting stents, prefilled inhalers, and syringes, as well as cellular analysis, nucleic, and blood glucose diagnostic devices. US disposable medical supplies was a $60.3 billion industry in 2010, with the largest players being Johnson & Johnson, Abbott Laboratories, Becton-Dickinson, Kimberly-Clark, Baxter Healthcare, 3M, Boston Scientific, and Roche.

The US medical device industry produces one-half of the world's medical devices and consumes approximately 40% of the world's output. Most of the market is held by just 17 companies that account for 65% of the total revenue in this segment. Although there are approximately 6,000 medical device manufacturers employing about 400,000 people, roughly 80% of those companies employ fewer than 50. Many are small, entrepreneurial firms that produce very specific devices. The market is projected to reach $68.9 billion this year and $74.5 billion in 2011.

7.1.2 High Tech IV Catheters

Medical device contract manufacturer NDH Medical, Inc. is patenting its novel extrusion process to produce unique catheter tubing for a new line of catheters.

D.V. Rosato, *Plastics End Use Applications*, SpringerBriefs in Materials,
DOI 10.1007/978-1-4614-0245-9_7, © Springer Science+Business Media, LLC 2011

The extrusion process produces multi-lumen catheter tubing with varying wall thicknesses. NDH used the new process in developing the first product utilizing the FlatCath Hemodialysis Catheter technology under an agreement with FlatCath, LLC of Georgia, a joint venture between Hatch Medical LLC and the tubing inventor, Aubrey Palestrant, M.D. Palestrant holds a patent on the tubing. Large, multi-lumen catheters are regularly placed into the venous system to perform hemodialysis, infusion or aspiration. Catheter placement often creates turbulence and sluggish blood flow resulting in a myriad of patient-related health concerns. The FlatCath technology facilitates a significantly reduced cross-sectional catheter profile within the blood vessel when the catheter is not in use reducing the risk of blood clots and complications. NDH Medical specializes in custom extrusion of single-lumen, paratubing, multi-lumen, and co-extruded tubing. The company provides insert molding, injection molding, pad printing, assembly, and other services as well as in-house tooling design and fabrication.

7.1.3   Life-Saving Items Safely Packaged

Storopack, a leading manufacturer of structural molded packaging, is using BASF's Neopor expandable PS for boxes used in the transport of blood products and organs. Since these life-saving products must be maintained at constant temperature, good thermal insulation is essential. Storopack molds Neopor, which comes in the form of small black beads, into silver-grey foam blocks and shapes. Neopor has outstanding heat insulation properties, high compressive strength, impact absorption, lightness and moisture resistance. It also contains infrared absorbers and reflectors, which lower the foam's thermal conductivity and give it its silver grey color.

One package is a blood product foam container called MonoTripleBox that Storopack designed with medical products company Delta T of Germany. The MonoTripleBox is a highly insulating Neopor box designed for shipping blood products by post or courier. The box, comprising two foam halves with walls 65–80 mm thick, can hold up to three temperature control packs, containing bags of blood or plasma. The temperature control packs by Delta T called 'x°-Control' resemble domestic freezer packs and contain a special hydrocarbon cooling medium, which is used to maintain the blood products at a stable temperature during transport. Once encased in Neopor, red blood corpuscles can be maintained at 2–8°C for over 24 h at an ambient temperature of 25°C.

7.1.4   Highly Maneuverable Surgical Microscope

Operating microscopes play a crucial role in the operating room. They must deliver top performance in a highly demanding environment and they must be lightweight, always accessible but never in the way during demanding and detailed work. Consequently, these instruments are equipped with highly sophisticated balancing systems that ensure high-precision movement of the complex optical system without manual involvement on the part of the surgeon. The arm which the microscope moves on must therefore be as light as possible. High-performance plastics like the lightweight, robust, Baydur 110 polyurethane system from Bayer MaterialScience AG are ideal for use in this application.

Leica utilizes Baydur 110 polyurethane for all six housing components in its new M520F40 operating microscope, manufactured by the Swiss polyurethane processor Puratech GmbH of Altendorf. The Baydur 110 molded parts have thin walls, making them lightweight while still extremely stiff. These properties enabled the Leica engineers to avoid using a sheet metal or aluminum framework, which would have made the arm significantly heavier. The Baydur material also has excellent flowability so that even complex product geometries such as ribbing, apertures, or undercuts can be realized easily in a single step. These thin-walled, high-strength housings allow complete and reliable protection of the electronics and delicate mechanical components. Also, as sharp corners and edges are avoided in the design, the operating microscope is easy to clean and disinfect. This is facilitated by the good coatability and excellent chemical resistance of the polyurethane.

7.1.5   Surgical Microtools

ERBE Elektromedizin GmbH has developed microtweezers for non-invasive surgery (keyhole surgery), with handle and working parts molded of glass-reinforced DuPont Zytel. This bipolar microtweezer is a multi-purpose instrument that can grasp, open, and prepare the tissue and administer low-intensity electrical current to stop blood flow. The handle's individual parts include the main housing, a cable channel at the back and a holder at the front for a metal tube, which guides wires that operate the tweezer jaws.

ERBE and system supplier PEZET chose Dupont's Zytel 70G30HSL in place of PEEK originally planned for the handle. The Zytel, a glass-reinforced, heat-stabilized, hydrolysis resistant grade of nylon 66, was modified by incorporation of a masterbatch to accept laser-marking. Very precise parts can be molded from this high-performance resin. Thanks to its dimensional stability, the parts retain their functionality, even after multiple sterilizations in an autoclave with superheated steam at 134°C. Like all nylons, it also offers long-term resistance to the cleaning agents used in hospitals and medical facilities, such as those involved in flushing the metal tube.

7.1.6   Micro-sized Assay Disks

Tecan has developed a fast, low cost, easy to use microlab, LabCD. As the microlab Uses less than 10 µL for analysis, it also cuts consumption of valuable test compounds by up to 90%. The system moves test compounds and reactants through tiny channels by spinning a CD-sized disk at precise speeds, to perform 48 simultaneous drug screening assays.

Scientists working in drug discovery, molecular biology, and related areas rely on high-throughput assays to evaluate how compounds are absorbed, metabolized, and excreted, and the toxic effects produced. The 5 mm thick, 124 mm diameter disk consists of an upper half containing the systems channels bonded to a lower half with the storage and measurement wells. Samples are transferred to the disk by Tecan's Genesis workstation and its Ultra system governs spinning and measures reaction products using absorbance, fluorescence, and luminescence. Tecan evaluated numerous transparent polymers for the disk before Ticona's Topas cyclic olefin copolymer (COC) was selected. The COC provides low autofluorescence and excellent UV transmittance in molded parts, minimizing background noise. Weidmann Plastics of Switzerland molds the disks. The firm's micromolding experience and proprietary technologies enabled it to meet the ±2 μm tolerances of the disk's tiny 50 μm channels. COC has the inherent dimensional stability necessary to hold those tolerances, and its surface energy allows the fluids to flow predictably into the microchannels.

7.1.7   Home Healthcare CPAP Systems

TAGA Medical Technologies is using a polycarbonate compound supplied by RTP Company to produce a proprietary set of baffles in its humidifier system used to help obstructive sleep apnea (OSA) patients. Known as the Velocity Passover humidifier, the system serves to counteract the dehydrating effect of use of Continuous Positive Air Pressure (CPAP), in which air is pumped to the back of the patient's throat to prevent the tissue collapse that blocks respiration in OSA patients. The humidifier provides cool, moist air to the patient during CPAP treatment minimizing the dehydrating effect that can lead to sinus problems.

TAGA Medical worked with RTP and molder Elyria Plastics to fabricate the plastic baffle system from a customized RTP 300 Series polycarbonate supplied as a transparent blue compound. As the humidifiers must be portable, strength and impact resistance are a must. Since walls are as thin as 0.050 in., the compound use required lubricity and superior flow properties. As the housing functions as a water reservoir, transparency allows the patient to see when a refill is required. The final product was formulated with FDA compliant resin and pigment, to meet Class 2 FDA 510K standards and incorporated high flow and lubrication additives. A notched Izod impact strength to 15 ft lbs/in. and flexural modulus of $0.34 \text{ psi} \times 10^6$ provided required strength and rigidity.

7.2   Emerging Applications

7.2.1   Emerging Applications Introduction

One of the most exciting emerging market in recent years involves conductive and light-emitting polymers which are changing the face of electronics. A movement is currently underway that will transform the way we view information, especially when we are on the move. Companies are now producing prototypes of ultra-thin, large-area, rollable displays on a routine basis. The technology will also make possible the production of much larger, 3-D high-definition displays for both indoor and outdoor applications. Engineers are also now capable of creating electroluminescent film that can be illuminated much like fireflies, with the application of a voltage. Elsewhere, many new applications are emerging in the medical area, with plastic bone a possible new market.

One strongest indication that a new technology has bridged the gap from an intriguing to emerging technology is when capital is being bet. One of the largest wagers made on new plastic is that placed by NatureWorks LLC on polylactides (PLA), a plastics made from a bio-based monomer created by fermenting corn. NatureWorks has invested $750 million and 10 years of R&D on their NatureWorks PLA plant, which recently began operation.

Another measure of an emerging market is high growth rate, typically at or above 10%. By this definition, plastic pallets not thought of as new are an emerging market. Global demand is still growing at better than 10%/year, and no single production process is yet predominant in their manufacture. Most applications on reaching their apex will have one process dominate, or two compete to some degree. For pallet processing, there are at least four plastics processes extensively used: injection molding, rotational molding, single- or twin-sheet thermoforming, and structural foam molding. Elsewhere, business is booming in blown film for stretch–hood applications. In Europe alone the stretch–hood web market is projected to grow by more than 22%/year through 2012, compared to 5.5%/year growth for stretch film.

7.2.2 Plastic Bone Wound Repair

Advanced Ceramics Research (ACR) Inc. has developed a new type of artificial bone strong enough to support new bone growth but porous enough to be absorbed and replaced by the human body. The material developed, dubbed 'Plasti-Bone,' is made from a biologically compatible plastic with a ceramic coating. The material which the company expects to reach market within 5 years holds promise over metallic inserts, which cannot be absorbed by the body and must eventually be replaced, and also biological materials that are too brittle to support any significant force.

In the case of a crushed arm bone for example, doctors could convert bone scans of the good arm into code for the construction of a replacement part for the injured arm bone. The code would be used by rapid prototyping equipment to make the desired artificial bone segment from a polymer coated with a thin layer of porous calcium phosphate. Natural bone begins bonding to the Plasti-Bone in 8 weeks. Eventually, the implant is 'bioresorbed' replaced by new bone tissue. ACR developed the material under a grant from the Office of Naval Research, which is seeking advanced bone treatments for the wounded. The ability to use rapid prototyping technology to quickly create custom fit artificial bone material is a major benefit of the product. The tissue engineering market is expected to reach $80 billion in sales by 2012.

7.2.3   Satellite TV Car Antennas

Sarnatech BNL, a specialist designer and manufacturer of innovative plastic bearing solutions, has teamed with KVH Industries to produce two molded parts for KVH's new TracVision A5 for cars. The system allows access to satellite TV while the vehicle is moving, using a roof-mounted pancake-shaped satellite TV antenna designed specifically for passenger vehicles. Motors rotate the antenna as the car turns. A computer control device keeps track of signal strength and rotates the antenna to keep it pointed at the satellite.

A bearing and a giant thrust race form the antenna's mobile base. With height of the low-profile antenna fixed at 4.5 in., an ultra thin bearing was required. Sarnatech designed a bearing 3/8 in. high using BASF's Ultraform POM (polyoxymethylene) copolymer, with the pitch circle of the raceway as large as possible to offer maximum support. Integrated features include belt drive teeth, brass inserts to affix it to the antenna unit, and molded pads for stability. Using plastic instead of steel cut the bearing weight by 50% and costs by a third. A giant thrust race with a 25 in. outside diameter was designed to support the perimeter of the 30.5 in. wide antenna. This large POM part was formed in eight identical sections with each section molded in one shot. The eight sections of the finished thrust race fit together effortlessly by means of an integral clip connection design.

7.2.4   Luminescent Handbags

Bree Collection GmbH, an international leather and bag specialist, has created an illuminated business handbag using Smart Surface Technology (SST). Bree had been toying with the idea of illuminating the dark insides of handbags for some time, but lacked an elegant solution. The answer is now available as Smart Surface Technology developed by Bayer Polymers, in partnership with Lumitec, a specialist in electroluminescence (EL) and precision electronic components. EL is a method of generating light reminiscent of the chemical method employed by fireflies. Engineers are using a film that lights up on application of a voltage to achieve electroluminescence. An advantage of EL is that it does not produce heat. However until now, only flat surfaces of limited size could be achieved. Smart Surface Technology makes it possible for the films to be shaped to illuminate any conceivable geometry. Incorporated in a nonconductive layer, the film in Bree's handbag lights up at the press of a button. Bayer Polymers, sees the main application for this technology in the auto industry. Incandescent lamps in cars will become obsolete. Instrument panels will be designed to take up less room and a car's interior headliner will glow in a soft glare free light.

7.2.5   *"Virtual Layering" Clothing Fabrics*

Malden Mills Industry's Polartec Heat Technology provides warmth on demand during stop and go activities. Battery powered "Polartec Heat" polyester fiber clothing panels reduce the need to add or subtract layers of clothing as weather or activity levels change. The thin, flexible, lightweight panels are durable, machine washable, and designed not to interfere with a garment's function. The heat controller with a molded PBT/ABS alloy housing provides temperature adjustment as necessary using three modes: full power, oscillate, or power off. The heat panels are powered by lightweight lithium ion batteries, rechargeable without removal from the garment, using standard 110 V household current, a solar powered charger, or a vehicle's 12 V cigarette lighter port. Battery life is 5 h on oscillate and 2.5 h on full power.

Thermal Mannequin Testing at the US Army Natick Soldier Systems Center have demonstrated the increase in thermal insulation (clo). When set on low, Polartec Heat Technology adds an additional .093 clo to the body's microclimate, the equivalent of adding an additional layer of Polartec 100 Series fleece. When set on high, an additional 1.51 clo is added to the body's microclimate, equivalent to adding an additional layer of Polartec 300 Series fleece. This "virtual layering" allows the wearer to quickly add and subtract insulation at the touch of a button.

7.3   Other Applications

7.3.1   *Government Applications Introduction*

Numerous US plastics injection molders have discovered that the best offense against global competition is a good defense – i.e., manufacturing for the largest purchaser of goods and services in the world, the US Dept. of Defense. The military budget for FY2011 is $600 billion with more than half that spent on hardware. Considering the homeland security implications, the military/defense market is one marketplace that is definitely not going offshore. Many new applications under development call for the lightweight functionality that plastic parts can provide and mirror many of the opportunities for plastics in the nonmilitary markets, such as aerospace, vehicular equipment and components, building and construction, E/E, telecom, medical/dental, appliances, packaging, and IT parts. In nonmilitary government markets, there is significant opportunity for plastics and plastic parts producers as much of the North American and European infrastructure, from bridges to tunnels, power and pipelines to telephone poles and street light standards and lighting systems, built in the 1950s and 1960s needs to be renovated, rebuilt, or replaced.

7.3.2   Driver Safe Lamp Posts

In response to European passive safety legislation (EN 12767, introduced to reduce deaths/serious injuries resulting from vehicles colliding with roadside structures) collapsing plastic light posts, able to absorb much of the force in a collision, have been developed as a Eureka project (a pan-European network for market-oriented, industrial R&D). There are approximately 165 million utility poles, and similar structures in use throughout Europe (16 million in the UK) with 200 motorist deaths and 3,000 injuries/year in the UK involving cars collisions with steel or concrete lighting columns. These columns and poles, in steel, aluminum, wood, or reinforced concrete, are prone to corrosion due to the combination of road salts, ground conditions, and dog's urine, limiting their life to 20–25 years and none behave in a passive safe manner.

'Thermopole' lamp post, a fiber reinforced plastics post, developed by Euro-Projects Ltd., is light weight, durable (40 years life), maintenance free, and is made of the only material to demonstrate passive safety performance when tested in accordance with the new European standard. FRP materials based on thermosetting polyester resins, used widely in flagpole construction, not been used in columns and signposts due to high cost, and lack of engineering confidence. Thermopole can withstand the elements, including high winds, but fails when a car hits it, crumbling or hinging to absorb the impact. Prototypes are stronger than steel, maintenance free, and recyclable. Over 400 composite lighting columns have been installed in the UK on a trial basis. To date there have been four major high-speed accidents involving the composite columns resulting in no deaths.

7.3.3   Ageless Underground Pipe

Dow's Continuum bimodal PE resins offer the opportunity to install a new, virtually leak-free pipe system or to cost-effectively rehabilitate an existing system using trenchless technology. The Continuum bimodal PE resins made by proprietary Unipol II process technology are a family of performance differentiated PE resins developed for the underground construction market. These PE resins provide longer life expectancy and outstanding strength for pressure pipe and fittings used in water and natural gas distribution, mining, and oil/gas processing. The resins offer improvements in long-term strength, temperature performance, and resistance to slow crack growth and rapid crack propagation compared to traditional PE pipe materials. The performance advantages of Continuum Bimodal PE resins provide an ideal solution for helping fix the nation's aging water, wastewater, and natural gas distribution infrastructure.

7.3.4 Super Strong Water Pipe

Sekisui Chemical has developed a new composite three-layer PE pipe that has superior strength to conventional PE water pipes, allowing significant reduction of wall thicknesses and raw material costs that more than offsets higher production costs. The company, which is exploring licensing the technology, supplies the pipes under the brand name Eslon PE Ultra. One area of focus for Sekisui's Urban Infrastructure and Environmental Products subsidiary that produces the new pipe is on new installations for water supply and sewer pipelines as well as replacements of conventional pipes with anti-corrosive piping.

The pipes consist of inner and outer layers of HDPE sandwiching a sheet of highly stretched (20–30 fold), highly oriented, spiral-wound Ecosurf HDPE sheet. The inner layer is first extruded, and the sheet then is wound round this inner layer. The semi-finished pipe is next fed through a special extrusion die to add the final layer. The sheet has a tensile modulus of 15 GPa, a tensile strength of 500 MPa, and elongation of 10%. The complex process is cost-effective for pipes with outside diameters of 200 mm, while 400 mm pipes are 30% lower in cost. A 200 mm Eslon PE Ultra pipe has half the wall thickness of Sekisui's single-layer PE pipes, allowing a single worker to carry a 5 m long pipe, easing installation. The first commercial application of the product was in high-pressure piping to pump water up a mountainside in Nagano Prefecture.

7.3.5 Composite UCAVs

The UCAV-N, unmanned combat air vehicle for the Navy, is envisioned as a ship-based, 'first day of the war' force multiplier. Designed with stealth features and shaped like a kite, Pegasus is constructed largely with composite materials. The aircraft measures 27.9 feet long and has a nearly equal wingspan of 27.8 feet. The goal of the joint DARPA/Navy project is to demonstrate the technical feasibility for a UCAV system to effectively and affordably conduct sea-based twenty-first century surveillance, suppression of enemy air defenses, and strike missions within the emerging global command and control architecture. Northrop Grumman's X-47A has successfully landed near a predesignated touchdown point to simulate the tail-hook arrestment point on a carrier flight deck. The shipboard-relative global positioning satellite system is used as the primary navigation source for increased landing precision.

7.3.6   Waste Management Applications Introduction

Since 1990, the plastics industry, as individual companies and through organizations such as the American Plastics Council, has invested more than $1 billion to support increased recycling and educate communities in the United States. Conserving resources means using less raw materials and energy throughout a product's entire life from development through disposal. Plastics are derived from natural resources – typically oil and natural gas. And yet, because of plastics' unique characteristics – lightweight, durability, formability – they often conserve more resources during a product's life when compared to other materials. While there has been much discussion in the US regarding waste management, recycle, and resource conservation, both Europe and Japan are global leaders in introducing legislation and implementing programs to control waste management and conserve resources.

7.3.7   Recycled PET Mine Shaft Reinforcement

Jennmar Corp., together with Terrasimco Inc., Napcor and the TOP Bottle Project have announced a major new technology and application use for post-consumer PET bottle material. The process developed by Terrasimco Inc., a Martinsburg WV technology company, has been licensed by Jennmar and will be used to produce new types of roof bolt products used to reinforce coal and other shaft mines. Pittsburgh-based Jennmar is the world's leading supplier of roof control products for the mining and tunneling industries. Jennmar has six manufacturing facilities in the US, two in Australia, and a new plant under construction in China. The new bolt will eliminate the two-step system previously used to install the bolt into the mine roof, by applying a polyester-based compound directly onto the bolt. The new product is called the 'Buddy Bar' to highlight the buddy system that is used in the mines during the installation of the reinforcing system.

Napcor (National Association for PET Container Resources) provided the expertise in identifying, developing, and sourcing appropriate post-consumer materials. The assistance provided by Napcor was part of the TOP Bottle Project initiated by Napcor to eliminate the barriers that prevent all PET bottles from being recycled irrespective of color or construction. Jennmar expects to have the first of five projected retrofits operational by the fourth quarter 2004. Demand for the post-consumer materials is anticipated to be as much as 15 million lb by the end of 2005.

7.3.8   Waste Carpet Fiber Products

Carpet recycling consortium CARE (Carpet America Recovery Effort) has established an economic way to pelletize unwashed, recycled nylon carpet. CARE, a jointly sponsored industry and government initiative, was established to increase the recycle/reuse of post-consumer carpet and reduce the disposal of waste carpet in landfills. More than 2.5 million tons of carpet are discarded each year and with landfill capacity declining, there is an environmental need to recycle/reuse carpet. As used carpet is heavy and bulky to handle, it is difficult and expensive to manage posing a burden on waste processors and often results in the illegal disposal of carpet by the general public.

CARE worked on the problem of used carpet disposal for 3 years with two engineering firms, Project Consultants & Associates Inc. of Florida, and Superior Polymer Systems Inc. of ON, Canada. The melt compounding and densification of shredded carpet proved too expensive. CARE found that a more viable solution was to 'cold' press the carpet waste in a pellet mill. The compressed pellets, 0.25–0.50 in. in diameter and 0.50–0.75 in. long, contain 46% nylon fiber, 35% calcium carbonate, 10% polypropylene, and 9% styrene-butadiene rubber. The pellets are reused by melt blending at 10–50% levels with a polyolefin. The nylon fibers, which do not melt in the polyolefin matrix, provide 85% of the properties of a comparable quantity of glass fiber of similar length. Carpet pellets have been tested as stiffeners for polyolefin railroad ties, marine pilings, and decking. Pellet test quantities are available from CARE.

7.3.9   Nylon Car Parts Recycling

DuPont Composite Recycle Technology, a closed-loop nylon recycling process, can convert parts made of glass or mineral filled nylon 6 or 66 into first-use quality material in a way that is economically viable and environmentally responsible. The new DuPont Composite Recycle Technology marks a giant step forward in helping automakers meet the European Union's mandates for end-of-life vehicle (ELV) closed-loop recycling. The EU directive requires that 85% by weight of a vehicle be capable of recovery and reuse by the end of 2005 (versus 75% currently). Results of 'in-use' testing of recycle content material generated by a development program with Denso Corp. (www.globaldenso.com) show Composite Recycle Technology (CRT) can be one of the most effective technologies.

The feedstock for the study consisted of 500 radiator end tanks collected from scrapped end-of-life vehicles in Japan. All of the tanks were made of glass reinforced nylon 66. The tanks were dirty, and the material had been degraded by years of contact with hot engine coolant. CRT dissolves the used polyamide, then filters away contaminants and fillers. The molecular weight of the recovered polyamide is increased to whatever level is the desired for the final application. The process generates resin equivalent to virgin nylon. The studies confirmed the process's capability to convert nylon auto parts from scrapped vehicles into glass reinforced nylon resins and parts with properties equivalent to those of virgin nylons for a workable, cradle-to-cradle solution of radiator end tanks. The process can likely be adapted to recycle air intake manifolds and other nylon parts.

7.3.10   Eco-efficient Mechanical Recycling

A study recently released by PlasticsEurope (formerly the Association of Plastics Manufacturers in Europe – APME) showed that mechanical recycling of plastic parts from cars is favorable from an eco-efficiency point-of-view only where large, easily accessible, monomaterial parts are concerned. Feedstock recycling and energy recovery are the most effective methods for recovering plastics in end-of-life cars in the majority of cases. The report provided the first quantitative data for the comparison of recycling and energy recovery options for plastics parts in end-of-life vehicles (ELV) and will be an important consideration in determining the fate of 8 million cars reaching the end of their useful lives each year in Europe. The use of plastics in cars is steadily growing and by 2015 it is expected that 1.3 million tonnes of plastics waste will come from 12 million end-of-life vehicles.

Six plastics waste management options (mechanical recycling, blast furnace, cement kiln, syngas production, waste combustion, landfill) were analyzed in the study with a focus on seven vehicle components (bumper, seat cushion, intake manifold, wash liquid tank, air duct, headlamp lens, mirror housing). Key findings of the study were (1) mechanical recycling is the method of choice mainly for large, easily accessible, mono material parts. (2) For most parts, feedstock recycling options and energy recovery are the methods of choice. (3) Whole life-cycle analysis shows that the use phase of plastics parts in cars that have the largest contribution to the environmental impacts. (4) Landfill shows the worst eco-efficiency of all the options studied.

7.3.11   Environmentally Marketed Irrigation Pipe

Polyethylene irrigation tubing is an inexpensive way to irrigate crops, but it is easily contaminated with soil and vegetation during the growing season and is difficult to reuse. The tubing, which can wind through miles of farmland, is typically used for only one season before it is replaced and creates a major waste stream for farms.

Polyethylene irrigation tubing manufacturer, Delta Plastics, has created a recycling system to collect the used material for recycle. Double G Farm of Arkansas is one operation that recycles its pipe to rid the farm of a waste stream. The 4,600 acre farm uses 120 rolls, or 30 miles of pipe a year. The PE tubing is easier to install, easier to pick back up, and is less expensive to handle than aluminum irrigation systems. Delta Plastics operates 104 collection sites traveling up to 300 miles at least twice a year to each site to pick up used pipe. Farmers who utilize the company's recycling service must buy their new pipe from Delta. The pipe is returned to the Delta Plastics plant in Stuttgart, AK, where it is cut, washed, and pelletized. Because of weathering's negative effect on the plastic's properties, the used material is not recycled into new irrigation pipe. Makers of trash bags and parking stops are the main consumers of the pellets. From 1998 through the end of 2002, Delta Plastics recycled 70 million lb of the irrigation tubing.

Chapter 8
Plastics End Use Applications Company Source Guide

Keywords Plastics • Metal • Rubber • Paper • Wood • Glass • Ceramic • Endusers • Fabricators • Suppliers

8.1 Plastic and Related Materials Segments

8.1.1 Source Guide Introduction

This source material covers the five primary competitive material areas and their interacting 20 major plastic end use market segments. Featured are 250 key market leader corporate sources, and 375 other major company sources. For organizational purposes the 25 major areas are arranged into five groups of five related segments, consisting of one group of competitive materials and four groups of end use markets.

Within each of the 25 segments there are two subdivisions which consist of ten key market leading corporations and 15 other major companies, organized alphabetically in each category, for a total of 625 entities.

The first subdivision, Primary, covers the top ten global starting points for company information in a given area of interest. The second subdivision, Other, parallels the preceding key database source approach from an identically patterned but more concise manner. Within each of the subdivisions, companies are qualified on the basis of specific segment Sales size, technical marketing product line breadth, global diversification, specialty niche positioning, and overall net Sales.

To facilitate cross comparison and review of company sources, the five tables (one for each of the five groups of five segments) summarize primary key market leading and other major companies across the twenty-five critical areas (five competitive materials, followed by twenty largest to smallest end use markets). Also in the tables, companies are identified as suppliers ("S"), fabricators ("F"), Endusers ("E"), or combinations thereof. Sales dollar figures are in $US for ease of comparison (B=billion). Company country of origin is indicated by two letter codes in parenthesis.

D.V. Rosato, *Plastics End Use Applications*, SpringerBriefs in Materials, DOI 10.1007/978-1-4614-0245-9_8, © Springer Science+Business Media, LLC 2011

8.1.2   Plastic, Metal, Paper & Wood, Rubber, Glass & Ceramic Companies

Table 8.1 Major endusers, fabicators and suppliers—Group One

Plastic	
Primary	Other
Asahi Kasei Corporation (JP); S www.asahi-kasei.co.jp Sales: $4.53 billion	Ashland Inc (US); S www.ashland.com
Basell N.V. (NL); S www.basell.com Sales: $5.7 billion	Chevron Phillips Chem. (US); S www.cpchem.com
	Daicel Chemical Ind. (JP); S www.daicel.co.jp/indexe.html
BASF AG (DE); S www.basf.com Sales: $7.2 billion	DSM N.V. (NL); S/F www.dsm.com
Bayer AG (DE); S www.bayer.com Sales: $11.6 billion	Ferro Corporation (US); S www.ferro.com
	Mannesmann Plastics M (DE); S www.mannesmann-plastics.com
Dow Chemical Co. (US); S/F www.dow.com Sales: $13.57 billion	Milicron Inc. (US); S www.milacron.com
Dupont Company (US); S/F www.dupont.com Sales: $18.4 billion	Polyone Corporation (US); S www.polyone.com
	Rohm & Haas Co. (US); S www.rohmhaas.com
Eastman Chemical Co (US); S www.eastman.com Sales: $2.6 billion	Sinochem Corporation (CN); S www.sinochem.com
Equistar Chem. LP (US); S www.equistarchem.com Sales: $5.54 billion	SMS Plastics Tech. (DE); S www.sms-k.com
	Solvay S.A. (BE); S www.solvay.com
Exxon Mobil Chemical (US); S www.exxonmobil.com/chemical Sales: $20.3 billion	Ticona Inc. (US); S www.ticona-us.com
G.E. Plastics (US); S/F www.geplastics.com Sales: $5.3 billion	Total Fina Elf-Atofina (FR); S/F www.atofina.com
	Wellman Inc. (US); S/F www.wellmaninc.com

Metal		Paper and wood
Primary	Other	Primary
Alcan Inc. (US); S/F www.alcan.com. Sales: $11.72 billion	AK Steel Corporation (US); S/F www.aksteel.com	Georgia Pacific (US); S/F www.gp.com. Sales: $18.95 billion
Alcoa Inc. (US); S/F www.alcoa.com Sales: $17.5 billion	ASARCO Inc. (MX); S/F www.asarco.com	International Paper (US); S/F www.internationalpaper.com Sales: $16.70 billion
	Bethlehem Steel (US); S/F www.bethsteel.com	
Arcelor (ES, LU, FR); S www.arcelor.com Sales: 28.6 billion	Commercial Metals (US); F www.commercialmetals.com	Kimberly-Clark (US); F www.kimberly-clark.com Sales: $14.5 billion
Corus Group plc (UK); S/F www.corusgroup.com Sales: $11.4 billion	INCO Ltd. (CA); S www.inco.com	MeadWestvaco Corp. (US); F www.meadwestvaco.com Sales est.: $6.86 billion
	International Steel Grp. (US); S/F www.ltvsteel.com	
Kobe Steel, Ltd. (JP); S/F www.kobelco.co.jp Sales: $6.28 billion	ISPAT Int'l N.V. (NL); S/F www.ispat.com	Nippon Paper Ind. Co. (JP); F www.npaper.co.jp Sales: $9.14 billion
Nippon Steel Corp. (JP); S/F www.nsc.co.jp Sales: $13.88 billion	Kawasaki Steel Corp. (JP); S/F www.kawasaki-steel.co.jp	OJI Paper Co. Ltd. (JP); S/F www.ojipaper.co.jp Sales: $7.90 billion
	Maxxam Inc. (US); S/F www.kaiseral.com	
NKK Corporation (JP); S/F www.nkk.co.jp Sales: $9.73 billion	Nucor Corporation (US); S/F www.nucor.com	SAPPI Ltd., (ZA); S/F www.sappi.com Sales: $3.73 billion
POSCO Ltd. (KR); S/F www.posco.co.kr Sales: $9.45 billion	Phelps Dodge Corp. (US); S/F www.phelpsdodge.com	Stora Enso OYJ (FI); S/F. www.storaenso.com Sales: $13.76 billion
	Ryerson Tull, Inc. (US); S/F www.ryersontull.com	
Rio Tinto plc (UK); S www.riotinto.com Sales: $10.83 billion	Sandvik AB (SE); S/F www.sandvik.com	UPM-Kymmmene. (FI); F www.upm-kymmene.com Sales: $8.0 billion
US Steel Corporation (US); S/F www.ussteel.com Sales: $6.95 billion	Teck Cominco Ltd. (CA); S www.teckcominco.com	Weyerhaeuser Co. (US); S/F. www.weyerhaeuser.com Sales: $15.1 billion
	Thyssen Krupp (DE); S/F www.thyssenkrupp.com	

(continued)

Table 8.1   (continued)

Paper and wood	Rubber	
Other	Primary	Other
Abitibi Consolidated (CA); S/F www.abicon.com	Bridgestone Corporation (JP); F www.bridgestone.co.jp Sales: $19.02 billion	Bandag Inc (US); S www.bandag.com
Aracruz Cellulose S.A. (BR); S/F www.aracruz.com.br	Michelin (FR); F www.michelin.com Sales: $11.62 billion	Bayer AG (DE); S www.bayerrubberone.com
Arjo Wiggins Appleton (UK); S/F www.awusa.com		Carlisle Co. Inc. (US); F www.carlisle.com
Boise Cascade Corp. (US); S/F www.bc.com	Continental AG (DE); F www.conti.de Sales: $5.48 billion	Dayco Products. Inc. (US); F www.dayco.com
Bowater Inc. (US); F www.bowater.com	Cooper Tire & Rubber (US); F www.coopertire.com Sales: $1.8 billion	Dow Corning Corp. (US); S www.dowcorning.com
Domtar Inc. (CA); S/F www.domtar.com		DSM N.V. (US); S www.dsmelastomers.com
Fletcher Challenge (NZ); S/F www.fcf.co.nz	Freudenberg & Company (DE); F www.freudenberg.com Sales: $4.32 billion	Dupont Dow Elastomers (US); S www.dupont-dow.com
Holmen AB (SE); S/F www.modogroup.com	Goodyear (US); S/F www.goodyear.com Sales: $13.0 billion	Exxon Mobil Chemical (US); S www.exxon.mobil.com
Louisiana-Pacific (US); S/F www.lpcorp.com		The Gates Corp. (UK); F www.gates.com
Nexfor Inc. (CA); S/F www.nexfor.com	Pirelli SpA (IT); F www.pirelli.com Sales: $3.08 billion	Gencorp Inc. (US); F www.gencorp.com/vs.html
Norske Skog. ASA (NO); S/F www.norske-skog.com	Sumitomo Rubber Ind. (JP); F www.sumitomorubber.co.jp Sales: $3.79 billion	Manuli Rubber Industries SpA www.manulirubber.com
Potlatch Corporation (US); S/F www.potlatchcorp.com		Sabanci Holding A.S. (TR); F www.sabanci.com.tr
Rayonier Inc. (US) S/F www.rayonier.com	Toyo Tire & Rubber (JP); S/F www.toyo-rubber.co.jp Sales: $1.91 billion	Sime Darby Berhad (MY); S/F www.simenet.com
Svenska Cellulosa (SE); F www.sca.se	Yokohama Rubber Co. (JP); F www.yrc.co.jp/home.html Sales: $3.01 billion	Vredestein N.V. (NL); S/F www.vredestein.com
Tembec Inc. (CA); S/F www.tembec.com		Waterville TG Inc. (JP); F www.wtg.ca

Glass and ceramic	
Primary	Other
Asahi Glass Company (JP); S/F www.agc.co.jp. Sales: $10.34 billion	Cemex, S.A. de C.V. (MX); S www.cemex.com
Carl-Zess-Stiftung (DE); F www.zeiss.com Sales: $2.43 billion	Cookson Group plc. (UK); S www.cooksongroup.co.uk
	Donnelly Corporation (US); F www.donnelly.com
Corning Inc. (US); S/F www.corning.com Sales: $3.16 billion	Engelhard Corporation (US); S www.engelhard.com
Elkem ASA (NO); S www.elkem.com Sales: $0.70 billion	Fisher Scientific Int'l Inc. (US); F www.fisherscientific.com
	Guardian Ind. Corp. (US); F www.guardian.com
Kyocera Corporation (JP); S/F www.kyocera.co.jp Sales: $8.76 billion	Harbison-Walker Refract. (US) S www.hwr.com
LaFarge S.A. (FR); S www.lafarge.fr Sales: $15.73 billion	Johns Manville Corp. (US); F www.jm.com
	Lancaster Colony Corp. (US); F www.glassonline.com
Owens Corning (US); S/F www.owenscorning.com Sales: $4.87 billion	Morgan Crucible Co. (UK); S www.morgancrucible.com
Pilkington plc (UK); F www.pilkington.com Sales: $4.56 billion	N.A. Refractories Co. (US); S/F www.hwr.com
	Seiko Epson Corporation (JP); F www.epson.co.jp
PPG Industries, Inc (US); S/F www.ppg.com Sales: $2.67 billion	Taiheiyo Cement Corp. (JP); S www.taiheiyo-cement.co.jp
Saint Gobain Group (FR); S/F www.saint-gobain.fr Sales: $20.48 billion	Tosoh Corporation (JP); F www.tosoh.com
	Waterford Wedgewood (IE); F tel: +353-147-8155

8.2   Major and Intermediate Plastic End Use Market Segments

8.2.1   *Major Packaging, Building & Construction, Automotive, Electrical & Electronic, and Appliance Companies*

Table 8.2   Major endusers, fabricators and suppliers—Group Two

Packaging	
Primary	Other
Ball Corporation (US); F www.ball.com Sales: $3.86 billion	Amcor Ltd. (AU); F www.amcor.com.au
Crown Cork & Seal Co. (US); F www.crowncork.com Sales: $6.8 billion	Bemis Company, Inc. (US); F www.bemis.com
	Dart Container Corp. (US); F www.dartcontainer.com
Groupo Vitro (MX); F www.vto.com Sales: $0.95 billion	Gerresheimer Group (DE); F www.gerresheimer.com
Owens-Illinois, Inc. (US); F www.o-i.com Sales: $5.64 billion	Greif Brothers Corp. (US); F www.greif.com
	Huhtamaki OYJ (FI); S/F www.huhtamaki.com
Pactiv Corporation (US); F www.pactiv.com Sales: $2.88 billion	Menasha Corp. (US); S/F www.menasha.com
Pechiney S.A. (FR); S/F www.pechiney.com Sales: $2.52 billion	Packaging Corp. (US); F www.packagingcorp.com
	Plastipak Packaging Inc. (US); F www.plastipak.com
Smurfit-Stone Container (US); F www.smurfit-stone.com Sales: $7.48 billion	Printpack, Inc. (US); F www.printpack.com
Sonoco Products Co (US); S/F www.sonoco.com Sales: $2.81 billion	Rexam plc (UK); F www.rexam.co.uk
	Sealed Air Corp. (US); F www.sealedaircorp.com
Temple Inland Inc. (US); S/F www.temple-inland.com Sales: $2.59 billion	Silgan Holdings, Inc. (US); F www.silgan.com
Toyo Seikan Kaisha (JP); S/F www.toyo-seikan.co.jp Sales: $3.49 billion	Solo Cup Company (US); F www.solocup.com
	Sweetheart Cup Co. (US); F www.sweetheart.com

Building and construction		Automotive
Primary	Other	Primary
American Standard Inc. (US); F www.americanstandard.com. Sales: $6.74 billion	Andersen Corporation (US); F www.andersencorp.com	DaimlerChrysler AG (DE); F/E www1.daimlerchrysler.com Sales: $145.5 billion
Daiwa House Industry (JP); F/E www.daiwahouse.co.jp Sales: $9.03 billion	AOKI Corporation (JP); F/E www.aoki.co.jp Armstrong World Ind. (US); F www.armstrong.com	Fiat Auto SPA (IT); F/E www.fiatgroup.com Sales: $34.98 billion
Kumagai Gumi Co. (JP); F/E www.kumagai.co.jp Sales: $3.83 billion	Centex Corporation (US); F/E www.centex.com	Ford Motor Company (US); F/E www.ford.com Sales: $134.4 billion
Masco Corporation (US); F www.masco.com Sales: $9.4 billion	Fujita Corporation (JP); F/E www.fujita.co.jp Haseko Corporation (JP); F/E www.haseko.co.jp	General Motors Corp. (US); F/E www.gm.com Sales: $148.0 billion
Mitsui Fudosan Co. (JP); F/E www.mitsuifudosan.co.jp Sales: $5.11 billion	D.R. Horton, Inc. (US); F/E www.drhorton.com	Honda Motor Co. (JP); F/E www.honda.com Sales: $55.25 billion
Obayashi Corporation (JP); F/E www.obayashi.co.jp Sales: $11.32 billion	KB Home (US); F/E www.kbhomes.com Kohler Company. (US); F www.kohlerco.com	Nissan Motor Co. (JP); F/E www.nissan-global.com Sales: $52.42 billion
Sekisui House, Ltd.(JP); F/E www.sekisuihouse.co.jp Sales: $11.00 billion	Lennar Corporation (US); F/E www.lennar.com	Peugeot Citroen S.A. (FR); F/E www.psa.fr Sales: $47.31 billion
Shimizu Corporation (JP); F/E www.shimz.co.jp Sales: $10.75 billion	Lennox International Inc. (US); F www.lennoxinternational.com Pulte Homes Inc. (US); F/E www.pulte.com	Renault S.A. (FR); F/E. www.renault.com Sales: $37.09 billion
Skanska AB (SE); F/E www.skanska.com Sales: $16.07 billion	Sherwin Williams Co. (US); S www.sherwin-williams.com	Toyota Motor Corp. (JP); F/E www.toyota.co.jp Sales: $114.05 billion
Taisei Corporation (JP); F/E www.taisei.co.jp Sales: $11.11 billion	USG Corporation (US); S/F www.usgcorp.com U.S. Industries, Inc. (US); F www.usindustries.com	Volkswagen AG (DE); F/E www.volkswagen.de Sales: $89.84 billion

(continued)

Table 8.2   (continued)

Automotive	Electrical and electronic	
Other	Primary	Other
ArvinMeritor, Inc. (US); F www.arvinmeritor.com	Fujitsu Ltd.. (JP); F/E www.fujitsu.com Sales: $18.64 billion	ABB Ltd. (CH); www.abb.com
BMW (DE); F/E www.bmw.com	Hewlett-Packard Co (US); F/E www.hp.com Sales: $45.96 billion	Alcatel SA (FR); F/E www.alcatel.com
Robert Bosch GmbH (DE); F www.bosch.de		Apple Computer Inc (US); F/E www.apple.com
Daihatsu Motor Corp. (JP); F/E www.daihatsu.com	Hitachi, Ltd. (JP); F/E www.hitachi.co.jp Sales: $24.94 billion	Cisco Systems Inc. (US); F/E www.cisco.com
Delphi Auto Systems (US); F www.delphiauto.com	IBM Corporation (US); F/E www.ibm.com Sales: $27.5 billion	Dell Computer Corp (US); F/E www.dell.com
Hyundai Motor Com. (KR); F/E www.hmc.co.kr		Emerson Electric Co (US); F/E www.gotoemerson.com
Isuzu Motors Ltd. (JP); F/E www.isuzu.co.jp	Lucent Technologies (US);F/E www.lucent.com Sales: $12.2 billion	General Electric Co. (US); F/E www.ge.com
Johnson Controls Inc. (US); F www.johnsoncontrols.com	Motorola, Inc. (US); F/E www.motorola.com $26.68 billion	Murata Mfg. Co. (JP); F/E www.murata.co.jp
Lear Corporation (US); F www.lear.com		Nokia Corporation (FI); F/E www.nokia.com
Magna International Inc. (CA); F www.magnaint.com	NEC Corporation. (JP); F/E www.nec.com Sales: $US 38.45 billion	Nortel Networks Corp. (CA); F/E www.nortel.com
Mazda Motor Corp. (US); F/E www.mazda.co.jp	Royal Philips Elect. NV (NL); F/E www.philips.com Sales: $31.12 billion	Sharp Corporation (JP) F/E www.sharp-world.com
Mitsubishi Motors Corp. (US) F/E www.mitsubishi-motors.co.jp		Sun Microsystems Inc. (US); F/E www.sun.com
Paccar Inc. (US); F/E www.paccar.com	Siemens AG (DE); F/E www.siemens.de Sales: $82.87 billion	Telefonaktiebolaget (SE); F/E www.ericsson.com
Visteon Corporation. (US); F www.visteon.com	Toshiba Corporation (JP); F/E www.toshiba.co.jp Sales: $30.68 billion	Tyco International Ltd. (BM); F/E www.tycoint.com
AB Volvo (SE); F/E www.volvo.com		Vishay Intertechnology (US); F/E www.vishay.com

Appliance	
Primary	Other
Bosch-Siemens (DE); F/E www.bsh-group.com Sales: $5.4 billion	Applica Inc. (US); F/E www.applicainc.com
Carrier Corporation (US); F/E www.global.carrier.com Sales: $8.77 billion	Elco Brandt S.A. (FR); F/E www.elcopourbrandt.com
	Fedders Corporation (US); F/E www.fedders.com
Electrolux, AB (SE); F/E www.electrolux.com Sales: $13.04 billion	Goodman Mfg. Co. (US); F/E www.goodmanmfg.com
G.E. Consumer Prod. (US); F/E www.geappliances.com Sales: $6.07 billion	Groupe SEB (FR); F/E www. groupeseb.com
	NACCO Housewares, (US); F/E www.hambeach.com
Maytag Corporation (US); F/E www.maytagcorp.com Sales: $4.66 billion	Juki Corporation (JP); F/E www.juki.co.jp
Miele & CIE, GmbH. (DE); F/E www.miele.de Sales: $2.42 billion	Lennox International (US); F/E www.lennoxinternational.com
	Melitta (DE); F/E www.melitta.de
Sunbeam Corp. (US); F/E www.sunbeam.com Sales: $1.7 billion	Merloni Elettrodomestici (IT); F/E www.merloni.com
The Trane Company (US); F/E www.trane.com Sales: $4.74 billion	Nortek Holdings Inc. (US); F/E www.nortek-inc.com
	The Rival Company (US); F/E www.rivco.com
Whirlpool Corporation (US); F/E www.whirlpoolcorp.com Sales: $11.02 billion	Salton Inc. (US); F/E www.saltoninc.com
York Int'l Corporation (US); F/E www.york.com Sales: $3.84 b illion	Singer NV (CN); F/E www.singerco.com
	Vorwerk and Company (DE); F/E www.vorwerk.de

8.2.2   *Intermediate Medical, Consumer Products, Toy, Recreation & Leisure, and Furniture Companies*

Table 8.3 Major endusers, fabricators and suppliers—Group Three

Medical	
Primary	Other
Abbott Laboratories (US); F/E www.abbott.com Sales: $5.88 billion	Bausch & Lomb Inc. (US); F/E www.bausch.com
	Biomet, Inc (US); F/E www.biomet.com
Baxter Int'l Inc. (US); F/E www.baxter.com Sales: $5.01 billion	
	CAPSUGEL (US); F www.capsugel.com
Becton Dickinson (US); F/E www.bd.com Sales: $2.15 billion	C.R. Bard, Inc. (US); F/E www.crbard.com
	Gambro AB (SE); F/E www.gambro.com
Boston Scientific (US); F/E www.bsci.com Sales: $2.92 billion	
	Hill-Rom Co. (US); F/E www.hillenbrand.com
GE Medical Systems (US);F/E www.gemedicalsystems.com Sales: $8.96 billion	Kendall Company, The (US); F/E www.kendallhq.com
	Medline Industries (US); F/E www.medline.com
Guidant Corporation (US); F/E www.guidant.com Sales: $3.24 billion	
	Omron Corporation (JP); F/E www.omronhealthcare.com
Hoffmann-La Roche (CH); F/E www.roche.com Sales: $5.31 billion	St Jude Medical (US); F/E www.sjm.com
	Smith & Nephew plc (UK); F/E www.smith-nephew.com
Johnson & Johnson (US); F/E www.jnj.com Sales: $19.15 billion	
	Stryker Corporation (US); F/E www.strykercorp.com
Medtronic, Inc. (US); F/E www.medtronic.com Sales: $6.41 billion	Sybron Dental (US); F/E www.sybrondental.com
	Wyeth (US); F/E www.wyeth.com
Siemens Medical (DE); F/E www.siemensmedical.com Sales: $8.20 billion	
	Zimmer Holdings, Inc. (US);F/E www.zimmer.com

Consumer products		Toy
Primary	Other	Primary
Eastman Kodak Co. (US); F/E www.kodak.com Sales: $6.52 billion	Bose Corporation (US); F/E www.bose.com	Bandai Co., Ltd. (JP); F/E www.bandai.com Sales: $1.65 billion
Gillette Co., The (US); F/E www.gillette.com Sales: $8.453 billion	Carrefour SA (FR); E www.carrefour.com	Hasbro, Inc. (US); F/E www.hasbro.com Sales: $2.86 billion
	Conair Corporation (US); F/E www.conair.com/corporate	
LG Electronics (KR); F/E www.lge.co.kr Sales: $6.88 billion	Daewoo Electronics (KR); F/E www.dwe.co.kr	K-B Toys (US); E www.kbtoys.com Sales: $2.0 billion
Matsushita Electric (JP); F/E www.panasonic.co.jp/global Sales: $11.88 billion	Fuji Photo Film Co. (JP); F/E www.fujifilm.co.jp	Lego Company (DK); F/E www.lego.com Sales: $1.1 billion
	Ito-Yokado Co., Ltd. (JP); E www.itoyokado.iyg.co.jp	
Newell Rubbermaid (US); F/E www.newell-rubbermaid.com Sales: $5.52 billion	Luxottica Group. SpA (IT); F/E www.luxottica.it/english	Little Tikes Co., The (US); F/E www.littletikes.com Sales: $0.45 billion
Pioneer Corporation (JP); F/E www.pioneer.co.jp Sales: $3.53 billion	Polaroid Corporation (US); F/E www.polaroid.com	Mattel, Inc. (US); F/E www.mattel.com Sales: $4.8 billion
	Samsung Electronics (KR); F/E www.samsungelectronics.com	
Sanyo Electric Co. (JP); F/E www.sanyo.co.jp Sales: $6.33 billion	Seiko Corporation (JP); F/E www.seiko-corp.co.jp	Nintendo Co. Ltd. (JP); F/E www.nintendo.com Sales: $3.5 billion
Sony Corporation (JP); F/E www.sony.co.jp/en/sonyInfo Sales: $17.16 billion	Snap-On Inc. (US); F/E www.snapon.com	SEGA Corporation. (JP); F/E www.sega.co.jp Sales: $1.82 billion
	TDK Corporation (JP); F/E www.tdk.com	
Stanley Works, The (US); F/E www.stanleyworks.com Sales: $1.955 billion	Tupperware Corp. (US); F/E www.tupperware.com	Toys "R" Us, Inc. (US); E www.tru.com Sales: $9.2 billion
3M Company (US); F/E www.mmm.com Sales: $2.792 billion	Wal-Mart Stores Inc. (US); E www.walmartstores.com	Ty Inc. (US); F/E www.ty.com Sales: $0.85 billion
	YKK Group (JP); F/E www.ykk.co.jp	

(continued)

Table 8.3 (continued)

Toy	Recreation and leisure	
Other	Primary	Other
Acclaim Entertainment (US); F/E www.acclaim.com	Adidas-Salomon AG (DE); F/E www.adidas-salomon.com Sales: $7.00 billion	AMF Bowling Inc. (US); F/E www.amf.com
Action Performance (US); F/E www.goracing.com	Bombardier Recreation (CA); F/E www.products.bombardier.com Sales: $1.70 billion	Callaway Golf Co. (US); F/E www.callawaygolf.com
Applause LLC (US); F/E www.applause.com		Coleman Co., The (US); F/E www.coleman.com
Capcom Co., Ltd. (JP); F/E www.capcom.co.jp	Brunswick Corporation (US); F/E www.brunswickcorp.com Sales: $3.712 billion	Head NV; F/E www.head.com
DSI Toys, Inc. (US); F/E www.dsitoys.com	Fleetwood Enterprises (US); F/E www.fleetwood.com Sales: $1.213 billion	Huffy Corporation (US); F/E www.huffy.com
Empire of Carolina Inc. (US); F/E www.empiretoys.com		Icon Health & Fitness (US); F/E www.iconfitness.com
First Years Inc., The (US); F/E www.thefirstyears.com	Fortune Brands (US); F/E www.fortunebrands.com Sales: $1.008 billion	Johnson Outdoors (US); F/E www.jwa.com
Flexible Flyer (US); F/E www.flexible-flyer.com	Harley-Davidson, Inc. (US); F/E www.harley-davidson.com Sales: $4.091 billion	K2 Inc. (US); F/E www.k2sport.com
Jakks Pacific, Inc. (US); F/E www.jakkspacific.com		Mizuno Corporation (JP); F/E www.mizuno.com
Marvel Enterprises Inc. (US); F/E www.marvel.com	Michaels Stores, Inc. (US); E www.michaels.com Sales: $2.856 billion	Recreational Equipment US); F/E www.rei.com
Midway Games Inc. (US); F/E www.midway.com	NIKE, Inc. (US); F/E www.nikebiz.com Sales: $9.893 billion	SCP Pool Corporation (US); F/E www.scppool.com
Playmates Toys Holding (CN); E www.playmatestoys.com		Shimano Inc. (JP) F/E www.shimano.com
RC2 Corporation (US) F/E www.rcertl.com	Reebok Int'l Ltd. (US); F/E www.reebok.com Sales: $3.128 billion	Skis Rossignol S.A. (FR); F/E www.skirossignol.com
Toy Quest (US) F/E www.manleytoyquest.com	Yamaha Corporation (JP); F/E www.yamaha.co.jp Sales: $3.79 billion	Spaulding Sports (US); F/E www.spalding.com
Wham-o, Inc. (US) F/E www.wham-o.com		Wilson Sport. Goods (US) F/E www.wilsonsports.com

Furniture	
Primary	Other
Furniture Brands Int'l (US); F/E www.furniturebrands.com. Sales: $2.398 billion	Ashley Furniture (US); F/E www.ashleyfurniture.com
Haworth Inc. (US); F/E www.haworth.com Sales: $1.66 billion	Bassett Furniture Ind. (US); F/E www.bassettfurniture.com
	Bush Industries Inc. (US); F/E www.bushfurniture.com
Herman Miller Inc. (US); F/E www.hermanmiller.com Sales: $1.469 billion	Chromcraft Revington (US); F/E Tel: (765) 564-3500
HON Industries Inc (US); F/E www.honi.com Sales: $1.692 billion	Craftmatic Ind. (US); F/E Tel: (215) 639-1310
	Dorel Industries, Inc. (CA); F/E www.dorel.com
IKEA International A/S (DK); E www.ikea.com Sales: $11.779 billion	Ethan Allen Inc. (US); F/E www.ethanallen.com
Kimball Int'l, Inc. (US); F/E www.kimball.com Sales: $ 0.736 billion	Hunter Douglas NV (NL); F/E www.hunterdouglas.com
	Klaussner Furniture (US); F/E www.klaussner.com
Knoll, Inc. (US); F/E www.knoll.com Sales: $0.964 billion	Leggett & Platt, Inc. (US); F/E www.leggett.com
LA-Z-Boy Inc. (US); F/E www.lazboy.com Sales: $2.154 billion	O'Sullivan Industries (US); F/E www.osullivan.com
	Pillowtex Corporation (US); F/E www.pillowtex.com
MFI Furniture Group (UK); F/E www.mfigroup.co.uk Sales: $2.065 billion	Sauder Woodworking (US); F/E www.sauder.com
Steelcase Inc. (US); F/E www.steelcase.com Sales: $ 2.600 billion	Serta Inc. (US); F/E www.serta.com
	Silentnight (UK); F/E www.silentnight-holdings.co.uk

8.3   Specialty and Niche Plastic Enduse Market Segments

8.3.1   *Specialty Office Products, Lawn & Garden, Marine & Boat, Aerospace, and Industrial Companies*

Table 8.4 Major endusers, fabricators and suppliers—Group Four

Office products	
Primary	Other
AVERY Dennison (US); F/E www.averydennison.com Sales: $1.812 billion	ACCO World Corp. (US); F/E www.acco.com
Canon, Inc. (JP); F/E www.canon.com Sales: $18.552 billion	Brother Industries Ltd. (JP); F/E www.brother.com
	Co. Machines Bull (FR); F/E www.bull.com
Lexmark Int'l Grp. (US); F/E www.lexmark.com Sales: $4.356 billion	Datacard Corporation (US) F/E www.datacard.com
Minolta Co., Ltd (JP); F/E www.minolta.com Sales: $2.88 billion	A.B. Dick Company (US); F/E www.abdick.com
	Diebold, Inc. (US); F/E www.dibold.com
NCR Corporation. (US); F/E www.ncr.com Sales: $2.963 billion	Esselte AB (SE); F/E www.esselte.com
Pitney Bowes Inc. (US); F/E www.pitneybowes.com Sales: $2.32 billion	Fellowes Manuf. Co (US); F/E www.fellowes.com
	IKON Office Sol'ns (US); E www.ikon.com
Ricoh Company Ltd. (JP); F/E www.ricoh.com Sales: $11.168 billion	Konica Corporation (JP); F/E www.konica.co.jp
Sharpie Group (US); F/E www.sanfordcorp.com Sales: $1.8 billion	Moore Corporation Ltd. (CA); F/E www.moore.com
	Océ N.V. (NL); F/E www.oce.com
Staples Inc. (US); E www.staples.com Sales: $11.596 billion	Palm Inc. (US); F/E www.palm.com
Xerox Corporation (US); F/E www.xerox.com Sales: $7.95 billion	Societe BIC S.A. (FR); F/E www.bic.fr
	United Stationers Inc. (US); E www.unitedstationers.com

Lawn and garden		Marine and boat
Primary	Other	Primary
Ames True Temper (US); F/E www.ames-truetemper.com Sales: $0.58 billion	Acorn Products (US); F/E www.uniontools.com	Alstom S.A. (FR); F/E www.alstom.com Sales: $1.081 billiom
Black/Decker Corp. (US); F/E www.bdk.com Sales: $0.98 billion	Agrium Inc. (CA); F/E www.agrium.com	Brunswick Corporation (US); F/E www.brunswickcorp.com Sales: $3.111 billion
	Blount Int'l, Inc. (US); F/E www.blount.com	
The Home Depot, (US); E www.homedepot.com Sales: $4.41 billion	Central Garden & Pet (US); E www.centralgardenandpet.com	DCN (FR); F/E www.dcnintl.com Sales: $1.61 billion
Husqvarna AB (SE); F/E www.international.husqvarna.com Sales: $1.98 billion	Davey Tree Expert, The (US); E www.davey.com	General Dynamics (US); F/E www.generaldynamics.com Sales: $3.650 billion
	Gilmour Manuf. Co (US); F/E www.gilmour.com	
LESCO, Ltd (US); F/E www.lesco.com Sales: $0.511 billion	Homebase Ltd. (UK); E www.homebase.co.uk	Genmar Holdings Inc. (US); F/E www.genmar.com Sales: $1.00 billion
MTD Products, Inc. (US); F/E www.mtdproducts.com Sales: $0.810 billion	Honda Power Equip. (JP); F/E www.hondapowerequipment.com	Hyundai Heavy Ind. (KR); F/E www.hhi.co.kr Sales: $3.208 billion
	Lawnware Products (US); F/E tel: (847) 966-3400	
Murray Inc. (US); F/E www.murrayinc.com Sales: $0.52 billion	Makita Corporation (JP); F/E www.makita.co.jp	Mitsubishi Heavy Ind. (JP); F/E www.mhi.co.jp Sales: $2.177 billion
Orkin Exterminating Co. (US); E www.orkin.com Sales: $0.665 billion	Sakata Seed Corp. (JP); F/E www..sakataseed.co.jp	Mitsui Eng. & Shipbuild. (JP); F/E www.mes.co.jp Sales: $1.008 billion
	Snapper Power Equip. (US); F/E www.snapper.com	
Scotts Company, The (US); F/E www.scottscompany.com Sales: $1.76 billion	Textron Golf and Turf (US); F/E www.ttcsp.com	Northrop Grumman (US); F/E www.nns.com Sales: $4.74 billion
Toro Company, The (US); F/E www.toro.com Sales: $1.399 billion	US Home & Garden Inc. (US); F www.ushg.com	Samsung Heavy Ind. (KR); F/E www.shi.samsung.co.kr Sales: $1.54 billion
	Woodstream Corporation (US); F www.woodstreamcorp.com	

(continued)

Table 8.4 (continued)

Marine and boat	Aerospace	
Other	Primary	Other
Aker Kværner Yards (NO); F/E www.kvaerner.com	Airbus S.A.S. (FR); F/E www.airbus.com Sales: $20.333 billion	Alcoa Aerospace. (US); F/E www.cordanttech.com
Boat America Corp. (US); E Tel: (703) 370-4202	Boeing Co., The (US); F/E www.boeing.com Sales: $28.387 billion	Arianespace SA (FR); F/E www.arianespace.com
Daewoo Shipbuilding. (KR); F/E www.daewooshipbuilding.com		Banner Aerospace Inc. (US); F/E www.banners.com
Davie Industries Inc. (CA); F/E www.davie.ca	Bombardier Aerospace (CA); F/E www.aero.bombardier.com Sales: $7.38 billion	Dassault Aviation Grp. (FR); F/E www.dassault-aviation.com
Dorbyl Marine Co. (ZA); F/E www.dorbylmarine.co.za	EADS (NL); F/E www.eads-nv.com Sales: $25.61 billion	Doncasters plc (UK); F/E www.doncasters.com
Fountain Powerboat (US); F/E www.fountainpowerboats.com		Embraer-Empresa (BR); F/E www.embraer.com
Greenway Partners, (US); F/E Tel: (212) 350-5100	G.E. Aircraft Eng. (US); F/E www.geae.com Sales: $11.141 billion	Fuji Heavy Industries (JP); F/E www.fhi.co.jp
Hobie Cat Company (US); F/E www.hobiecat.com	NASA (US); E www.nasa.gov Sales: $14.035 billion	Goodrich Corp. (US); F/E www.goodrich.com
Hunter Marine Corp (US); F/E www.huntermarine.com		Hexel Corporation (US); F/E www.hexcel.com
J. Ray McDermott (US) F/E www.jraymcdermott.com	Pratt & Whitney (US); F/E www.pratt-whitney.com Sales: $7.645 billion	Israel Aircraft Industries (IL); F/E www.iai.co.il
Kawasaki Heavy Ind. (JP);F/E www.khi.co.jp	Rolls-Royce plc (UK); F/E www.rolls-royce.com Sales: $6.525 billion	Kaman Corporation (US) F/E www.kaman.com
Oceaneering Int'l. (US); F/E www.oceaneering.com		Loral Space & Com. (US); F/E www.loral.com
Oshima Shipbuilding (JP);F/E www.osy.co.jp	SNECMA (FR); F/E www.snecma.com Sales: $7.002 billion	Magellan Aerospace (CA);F/E www.malaero.com
Universal Shipbuilding (JP); F/E www.hitachizosen.co.jp	Textron Inc. (US); F/E www.textron.com Sales: $4.922 billion	Orbital Sciences (US); F/E www.orbital.com
Viking Yacht Co. (SE); F/E www.vikingyachts.com		Smiths Group plc (UK); F/E www.smiths-ind-aerospace.com

Industrial	
Primary	Other
ABB Ltd. (CH); F/E www.abb.com. Sales: $21.7 billion	Briggs & Stratton Corp. (US); F/E www.briggsandstratton.com
Caterpillar, Inc. (US); F/E www.cat.com Sales: $18.602 billion	Eaton Corporation (US); F/E www.eaton.com
	Fanuc Ltd. (JP); F/E www.fanuc.com
G.E. Industrial (US); F/E www.ge.com Sales: $35.00 billion	Halliburton Co. (US); F/E www.halliburton.com
Illinois Tool Works (US); F/E www.itwinc.com Sales: $9.22 billion	ITT Industries, Inc. (US); F/E www.ittind.com
	Kawasaki Heavy. Ind. (JP); F/E www.khi.co.jp
Ingersoll-Rand Co. (US); F/E www.irco.com Sales: $9.51 billion	Komatsu Ltd. (JP); F/E www.komatsu.com
Kubota Corporation (JP); F/E www.kubota.co.jp Sales: $6.51 billion	NACCO Materials Hd (US); F/E www.nacco.com
	NSK Ltd. (JP); F/E www.nsk.com
Man AG (DE); F/E www.man.de Sales: $6.85 billion	Parker Hannifin Corp. (US); F/E www.parker.com
Mitsubishi Electric (JP); F/E www.mitsubishielectric.com Sales: $17.74 billion	Pentair, Inc. (US); F/E www.pentair.com
	Rockwell Automation. (US); F/E www.rockwellautomation.com
Mitsubishi Heavy. Ind. (JP); F/E www.mhi.co.jp Sales: $15.21 billion	Schlumberger Ltd. (FR); F/E www.slb.com
Schneider Electric S.A (FR); F/E www.schneiderelectric.com Sales: $ 9.754 billion	Sulzer Ltd. (CH); F/E www.sulzer.com
	Timken Company, The (US); F/E www.timken.com

8.3.2   Niche Agriculture, Waste Management, Government, Export, and Other & Emerging Companies

Table 8.5   Major endusers, fabricators and suppliers—Group Five

Agriculture	
Primary	Other
Archer Daniels Midland (US); E www.admworld.com Sales: $23.454 billion	AGCO Corporation (US); F/E www.agway.com
Cargill Inc. (US); E www.cargill.com Sales: $41.1 billion	Agway, Inc. (US); E www.agway.com
	Campbell Soup Co. (US); E www.campbellsoups.com
Conagra, Inc. (US); E www.conagra.com Sales: $27.63 billion	Chiquita Brands Int'l (US); E www.chiquita.com
Danone Group (FR); E www.danonegroup.com Sales: $14.237 billion	CNH Global N.V. (US); F/E www.cnh.com
	Deere & Company. (US); F/E www.deere.com
General Mills, Inc. (US); E www.generalmills.com Sales: $7.949 billion	Dole Food Company (US); E www.dole.com
Kraft Foods, Inc. (US); E www.kraft.com Sales: $29.723 billion	Farmland Industries. (US); E www.farmland.com
	Frito-Lay Company. (US); E www.fritolay.com
McDonald's Corp. (US); E www.mcdonalds.com/corporate Sales: $15.406 billion	H.J. Heinz Company (US); E www.heinz.com
Nestlé S.A. (CH); E www.nestle.com Sales: $60.559 billion	Hormel Foods Corp. (US); E www.hormel.com
	Kellogg Company (US); E www.kelloggs.com
Sysco Corporation (US); E www.sysco.com Sales: $19.163 billion	McCormick & Company (US); E www.mccormick.com
Unilever plc (UK); E www.unilever.com Sales: $28.48 billion	Sara Lee Corporation (US); E www.saralee.com
	Tyson Foods, Inc. (US); E www.tyson.com

Waste management		Government
Primary	Other	Primary
Allied Waste Ind. (US); E www.alliedwaste.com Sales: $5.517 billion	Brambles Industries Ltd. (AU); E www.brambles.com	BAE Systems plc (UK); F/E www.baesystems.com Sales: $14.861 billion
Danaher Corporation (US); F/E www.danaher.com Sales: $3.385 billion	Calgon Carbon Corp. (US); F/E www.calgoncarbon.com Casella Waste Systems (US); E www.casella.com	Bechtel Group, Inc (US); F/E www.bechtel.com Sales: $8.52 billion
Dover Industries (US); F/E www.dovercorporation.com Sales: $1.124 billion	Clean Harbors, Inc. (US); E www.cleanharbors.com Compaction America (US); F/E www.uniteddominion.com	Boeing Integ. Defense (US); F/E www.boeing.com/ids Sales: $19.97 billion
Ebara Corporation (JP); F/E www.ebara.co.jp Sales: $2.21 billion	Crane Environmental (US); F/E www.cranenv.com	Bouygues SA (FR); F/E www.bouygues.fr Sales: $15.186 billion
Philip Services Corp. (CA); E www.contactpsc.com Sales: $1.119 billion	ELG Haniel GmbH (DE); E www.haniel.de, www.elg.de GE Osmonics, Inc. (US); F/E www.osmonics.com	Fluor Corporation (US); F/E www.fluor.com Sales: $8.30 billion
Republic Services (US); E www.republicservices.com Sales: $2.365 billion	Ionics Inc. (US); F/E www.ionics.com	General Dynamics (US); F/E www.generaldynamics.com Sales: $12.45 billion
Safety-Kleen Corp. (US); E www.safety-kleen.com Sales: $1.515 billion	Kelda Group plc (UK); E www.keldagroup.com Norcal Waste Systems (US); E www.norcalwaste.com	Kajima Corporation (JP); F/E www.kajima.com Sales: $11.14 billion
Suez SA (FR); F/E www.suez.fr Sales: $4.64 billion	Shaw E & I Inc. (US); E www.shaw.grp.com	Lockheed Martin Corp. (US); F/E www.lockheedmartin.com Sales: $23.92 billion
Veolia Environment (FR); F/E www.vivendienvironnement.com Sales: $6.60 billion	Sumitomo Heavy Ind. (JP); F/E www.shi.co.jp Tetra Tech, Inc. (US); E www.tetratech.com	Northrop Grumman (US); F/E www.northgrum.com Sales: $16.0 billion
Waste Management (US); E www.wm.com Sales: $11.142 billion	Waste Industries Inc. (US); E www.waste-ind.com	Raytheon Company (US); F/E. www.raytheon.com Sales: $11.73 billion

(continued)

Table 8.5 (continued)

Government	Export	
Other	Primary	Other
Alliant Techsystems (US); F/E www.atk.com	ABB Ltd. (CH); F/E www.abb.com. Sales: $21.7 billion	Alcoa Inc. (US); S/F www.alcoa.com
EADS D&SS (DE); F/E www.eads-nv.com	Boeing Co., The (US); F/E www.boeing.com Sales: $54.069 billion	Bridgestone Corporation (JP); F www.bridgestone.co.jp
Foster Wheeler (US); F/E www.fwc.com		Brunswick Corporation (US); F/E www.brunswickcorp.com
GKN plc (UK); E www.gknplc.com	Du Pont Company (US); S/F www.dupont.com Sales: $18.4 billion	Canon, Inc. (JP); F/E www.canon.com
HBG NV (NL); F/E www.hbg.nl	General Electric (US); S/F/E www.ge.com Sales: $130.685 billion	Crown Cork and Seal (US); F www.crowncork.com
Hochtief AG (DE); F/E www.hochtief.de		Electrolux, AB (SE); F/E www.electrolux.com
Peter Kiewit Sons Inc. (US); F/E www.kiewit.com	General Motors (US); F/E www.gm.com Sales: $148.0 billion	Johnson & Johnson (US); F/E www.jnj.com
Philipp Holzmann AG (DE); F/E www.philipp-holzmann.de	International Paper (US); S/F www.internationalpaper.com Sales: $16.70 billion	Lockheed Martin Corp. (US); F/E www.lockheedmartin.com
Rosoboronexport (RU); E www.rusarm.ru		Mattel, Inc. (US); F/E www.mattel.com
SAIC (US) F/E www.saic.com	Mitsui & Co., Ltd. (IT); S/F/E www.mitsui.co.jp Sales: $95.146 billion	NIKE, Inc. (US); F/E www.nikebiz.com
Thales SA (FR); F/E www.thalesgroup.com	Nestlé S.A. (CH); E www.nestle.com Sales: $60.559 billion	Saint Gobain Group (FR) S/F www.saint-gobain.fr
TRW, Inc. (US); F/E www.trw.com		Steelcase Inc. (US); F/E www.steelcase.com
United Defense Ind. (US); F/E www.uniteddefense.com	Siemens AG (DE); F/E www.siemens.de Sales: $82.87 billion	Taisei Corporation (JP); F/E www.taisei.co.jp
Vinci SA (FR); F/E www.groupe-vinci.com	Sony Corporation (JP); F/E www.sony.co.jp Sales: $17.16 billion	Toro Company, The (US); F/E www.toro.com
Washington Group Int'l (US); F/E www.wgint.com		Waste Management Inc. (US); E www.wm.com

Other and emerging	
Primary	Other
Armor Holdings, Inc. (US); F/E www.armorholdings.com Sales: $0.305 billion	Caliper Technologies (US);F/E www.calipertech.com
Friede Goldman H. (US); F/E www.fgh.com Sales: $0.246 billion	Cree, Inc. (US); F/E www.cree.com
	Digital Lightwave Inc. (US); F/E www.lightwave.com
Hain Celestial Grp., The (US); E www.hain-celestial.com Sales: $0.396 billion	Fiskars Corporation (FI); F/E www.fiskars.fi
Industrial Distribution (US); E www.idglink.com Sales: $0.493 billion	Hi-Rise Recycling Inc. (US); F/E www.hiriserecycling.com
	Interiors, Inc. (US); F/E www.interiors.com
Kellstrom Industries (US); F/E www.kellstrom.com Sales: $0.354 billion	Jakks Pacific, Inc. (US); F/E www.jakkspacific.com
NCI Building Systems (US); F/E www.ncilp.com Sales: $0.953 billion	K-Swiss Inc. US); F/E www.kswiss.com
	Landec Corporation (US); F/E www.landec.com
Salton Inc. (US); F/E www.saltoninc.com Sales: $0.923 billion	Matrixx Initiatives Inc., (US); F/E www.gum-tech.com
Scansource, Inc. (US); E www.scansource.com Sales: $0.842 billion	Polymedica Corp. (US); E www.polymedica.com
	Standard Auto. Corp. (US); F/E www.standardauto.com
SLI, Inc. (US); F/E www.sli-lighting.com Sales: $0.851 billion	Summa Industries (US); F/E www.summaindustries.com
Solectron, Inc. (US); F/E www.solectron.com Sales: $12.27 billion	Thermoview Industries (US); F/E www.thermoviewinc.com
	US Plastic Lumber (US); F/E www.usplasticlumber.com

About the Author

Dr. Donald V. Rosato, President of PlastiSource, Inc. a prototype manufacturing, technology development, and marketing advisory firm in Massachusetts, USA is internationally recognized as a leader in plastics technology, business, and marketing. He has extensive technical and marketing, plastic industry business experience from laboratory testing, through production to marketing, having worked for Northrop Grumman, Owens-Illinois, DuPont/Conoco, Hoechst Celanese/Ticona, and Borg Warner/G.E. Plastics. He has developed numerous polymer related patents and is a participating member of many trade and industry groups. Relying on his unrivaled knowledge of the industry plus high level international contacts, Dr. Rosato is also uniquely positioned to provide an expert, inside view of a range of advanced plastics materials, processes and applications through a series of seminars and webinars. Among his many accolades, Dr. Rosato has been named Engineer of the Year by the Society of Plastics Engineers. Dr. Rosato has written extensively, authoring or editing numerous papers, including articles published in the Encyclopedia of Polymer Science and Engineering, and major books, including the Concise Encyclopedia of Plastics, Injection Molding Handbook 3rd Ed., Plastic Product Material and Process Selection Handbook, Designing with Plastics and Advanced Composites, and Plastics Institute of America Plastics Engineering, Manufacturing and Data Handbook. Dr. Rosato holds a BS in Chemistry from Boston College, MBA at Northeastern University, M.S. Plastics Engineering from University of Massachusetts Lowell, and Ph.D. Business Administration at University of California, Berkeley.

D.V. Rosato, *Plastics End Use Applications*, SpringerBriefs in Materials,
DOI 10.1007/978-1-4614-0245-9, © Springer Science+Business Media, LLC 2011